Lecture Notes in Mathematics

Volume 2305

This series reports on new developments in all areas of mathematics and their applications - quickly, informally and at a high level. Mathematical texts analysing new developments in modelling and numerical simulation are welcome. The type of material considered for publication includes:

1. Research monographs
2. Lectures on a new field or presentations of a new angle in a classical field
3. Summer schools and intensive courses on topics of current research.

Texts which are out of print but still in demand may also be considered if they fall within these categories. The timeliness of a manuscript is sometimes more important than its form, which may be preliminary or tentative.

Titles from this series are indexed by Scopus, Web of Science, Mathematical Reviews, and zbMATH.

Fabio Nicola • S. Ivan Trapasso

Wave Packet Analysis of Feynman Path Integrals

 Springer

Fabio Nicola
Department of Mathematical Sciences
Politecnico di Torino
Torino, Italy

S. Ivan Trapasso
MaLGa Center - Department of
Mathematics
University of Genova
Genova, Italy

ISSN 0075-8434 ISSN 1617-9692 (electronic)
Lecture Notes in Mathematics
ISBN 978-3-031-06185-1 ISBN 978-3-031-06186-8 (eBook)
https://doi.org/10.1007/978-3-031-06186-8

Mathematics Subject Classification: 81S40, 81S30, 81Q30, 81Q05, 35S05, 35S30, 42B35, 42B10, 47L80, 46T12

This Springer imprint is published by the registered company Springer Nature Switzerland AG
The registered company address is: Gewerbestrasse 11, 6330 Cham, Switzerland

Dedicated to our families, for their continuous support

Preface

This book provides an accessible and essentially self-contained presentation of some mathematical aspects of Feynman path integrals in non-relativistic quantum mechanics. In spite of the primary role in the advancement of modern theoretical physics and the wide range of applications, path integrals are still a source of challenging problems for mathematicians. From this standpoint, path integrals can be roughly viewed as approximation formulas for an operator (usually the propagator of a Schrödinger-type evolution equation) involving a suitably designed sequence of operators and different levels of convergence.

There are many excellent treatises on the mathematics of path integrals, regarded from different angles and for different approximation schemes [5, 39, 111, 154, 157, 174, 208], but the results contained in this book have never appeared before in a book. Indeed, they reflect the topics addressed in the research activity of the authors over the last years. In keeping with the spirit of harmonic analysis, we focus on the issue of the pointwise convergence of the integral kernels of the approximation operators, as well as the convergence in the spaces of bounded operators on L^2 and on L^p-based Sobolev spaces, $1 < p < \infty$, with optimal loss of derivatives. In addition to the case of smooth potentials, we consider low-regular potentials in Kato-Sobolev spaces or in the so-called Sjöstrand class, which contains the class of potentials considered by S. Albeverio et al. [1, 2, 4, 5] and K. Itô [147], namely those coinciding with the Fourier transform of a complex (finite) measure.

In short, the analysis is based on the decomposition of functions and operators by means of the so-called Gabor wave packets, using a machinery developed mainly (but not only) by H. Feichtinger and K. Gröchenig and collaborators over the last 30 years in the context of time-frequency analysis and, with a different terminology, by several authors working on partial differential equations—especially A. Boulkhemair, C. Fefferman, J. Sjöstrand, and D. Tataru. This framework allows one to lift the original problem from the configuration space to the phase space, the subtle oscillation issues being embedded into this lifting procedure. The outcome of this approach involves classes of integral operators in phase space –with continuous and band-dominated kernel– which turn out to be much more tamable than the original pseudodifferential and (global) Fourier integral operators with regard to boundedness, composition, and inversion properties.

These techniques have shown to be surprisingly successful in dealing with path integrals. For instance, they were recently used in [183] to prove the *pointwise convergence of the integral kernels* –in quantum parlance, the transition amplitudes in the position representation– in the Feynman–Trotter formula for low-regular potentials in the Sjöstrand class. This is a longstanding problem that was actually raised by Feynman himself [102, 104], but so far, a positive answer was given only for certain classes of *smooth* potentials by D. Fujiwara and collaborators in highly technical works [109, 111]. In Chap. 6, we will discuss the above-mentioned result, which is moreover global in time (except for a certain discrete subset of exceptional times). We will also consider even more general low-regular, possibly non-local, potentials.

Besides the specific topic, which seems worthy of further exploration, this book shall be intended as an attempt to build a bridge between the communities of people working in time-frequency analysis and mathematical/theoretical physics, as well as to provide an exposition of the present novel approach along with its basic toolkit. Having in mind a researcher or a Ph.D. student as a reader, we collected in Part I the necessary background, in the most suitable form for our purposes, following a smooth pedagogical pattern. In particular, proofs of preliminary results are given whenever they are instructive to make the reader confident enough with the underlying techniques. Then, Part II covers the analysis of path integrals with the above-mentioned convergence results. The book can also be used for a Ph.D. (or self-study) course, covering, for example, Part I (except for the sections "Complements") and Chaps. 6 and 7. For the benefit of the reader, the Index of Notation is divided into categories, with a short explanation next to most entries, while the bibliographical entries are organized using the alphabetic BibLaTeX style.

We would like to take this opportunity to warmly thank several colleagues and friends for uncountable inspiring discussions during a longstanding collaboration, in particular Elena Cordero, Hans Feichtinger, Maurice de Gosson, Karlheinz Gröchenig, and Luigi Rodino. We also wish to express our sincere gratitude to Sergio Albeverio, Maurice de Gosson (again) and Franz Luef for carefully reading part of the material contained in this monograph and to Riccardo Adami, Luigi Ambrosio, Ernesto De Vito, Daisuke Fujiwara, Massimiliano Gubinelli, Naoto Kumano-go, Sonia Mazzucchi, Luigi Ettore Picasso, Nicola Pinamonti, Paolo Tilli, Stefan Waldmann, and Kenji Yajima for discussions or correspondence on some aspects of this topic.

Torino, Italy   Fabio Nicola
Genova, Italy   S. Ivan Trapasso
March 2022

Contents

Outline

We briefly describe the organization of the material in this book.

First, in Chap. 1, the reader can find an itinerary through the topics of the book, which is an essential introduction to Gabor analysis and mathematical path integrals. We try to outline the key points of wave packet analysis and to describe the state of the art in rigorous Feynman path integrals from an operator-theoretic point of view, without any claim of completeness but only for motivational purposes. Indeed, we provide a description of the problems that we have taken into account and an exposition of the main results that will be proved.

The rest of the book is organized in two parts:

- Part I comprises the background material on time-frequency analysis. In an attempt to provide a self-contained presentation, in Chap. 2, we fix the notation and recall the general notions of analysis that are needed below, whereas Chaps. 3 and 4 are devoted to the basic results of Gabor analysis of functions/distributions and operators, respectively. In Chap. 5, we also provide a reformulation of this material in the spirit of semiclassical analysis.
- In Part II, we study the problem of the convergence for Feynman path integrals. More in detail:

  - The problem of pointwise convergence at the level of integral kernels for Feynman-Trotter approximate propagators is the focus of Chap. 6.
  - In Chap. 7, we deal with the problem of convergence in the norm operator topology in L^2 for a certain family of rough parametrices and for a class of potentials with low regularity.
  - Chapter 8 is devoted again to convergence results in the norm operator topology in L^2 for a refined class of parametrices and for a class of potentials with low Sobolev-type regularity.
  - Finally, in Chap. 9, we are confronted with convergence results in the same spirit as above in the L^p setting, $1 < p < \infty$, characterized by an unavoidable loss of derivatives.

Chapter 1
Itinerary: How Gabor Analysis Met Feynman Path Integrals

This monograph deals with several aspects of a problem which lies at the interface between analysis and physics, namely path integrals, where recourse to techniques of phase space analysis is particularly well suited.

1.1 The Elements of Gabor Analysis

To be precise, the ensemble of techniques used in this book should be referred to as *time-frequency analysis* or, even better, *Gabor analysis*. The origins of this fascinating branch of modern harmonic analysis date back to D. Gabor's article *Theory of Communication* of 1946 [113], where the author suggested that a family of functions obtained by translation and modulation of a single Gaussian function may provide a collection of elementary building blocks (usually known as atoms, wave packets or coherent states depending on the context) for any square-integrable signal. In detail, this means that for any $f \in L^2(\mathbb{R})$ there exist coefficients $c_{mn} \in \mathbb{C}$, $m, n \in \mathbb{Z}$, such that

$$f(x) = \sum_{m,n \in \mathbb{Z}} c_{mn} e^{2\pi i n x} g(x - m), \quad g(x) = e^{-\pi x^2}. \tag{1.1}$$

The same problem can be considered in the general d-dimensional Euclidean setting and also with different atoms than Gaussian ones, provided that suitable decay and smoothness conditions are guaranteed—for instance, one may assume $g \in \mathcal{S}(\mathbb{R}^d)$, the Schwartz class of rapidly decreasing functions.

Interestingly enough, Gaussian wave packets were already well known in physics since the early work of E. Schrödinger on minimum uncertainty states [200] and also in connection with coherent states of the Weyl-Heisenberg group [116, 188] and von Neumann lattices [229]. In passing, it was J. von Neumann the one

© The Author(s), under exclusive license to Springer Nature Switzerland AG 2022
F. Nicola, S. I. Trapasso, *Wave Packet Analysis of Feynman Path Integrals*,
Lecture Notes in Mathematics 2305, https://doi.org/10.1007/978-3-031-06186-8_1

who claimed (without proof) in [229] that the family of Gaussian functions $G = \{e^{2\pi i n x} e^{-\pi (x-m)^2}\}_{m,n \in \mathbb{Z}}$ spans a dense subset of $L^2(\mathbb{R})$.

Both the conjectures by Gabor and von Neumann turned out to be true, but proofs were given only in the 1970s [10, 187]. Nevertheless, the expansion (1.1) is unstable in many aspects: for instance, even for $f \in \mathcal{S}(\mathbb{R})$ the series in (1.1) converges only in the sense of distributions [149]. Moreover, the sequence of coefficients (c_{mn}) is not uniquely determined and does not precisely characterize the signal f in terms of energy, meaning that $\|c_{mn}\|_{\ell^2(\mathbb{Z}^2)}$ is not comparable in general with $\|f\|_{L^2(\mathbb{R})}$. In more precise terms, G is not a *frame* nor a *Riesz basis* for $L^2(\mathbb{R})$ [42, 134]. In fact, the main obstruction here resides in the choice of a critical time-frequency density: it is now well known that the family $\{e^{2\pi i \beta n x} e^{-\pi (x-\alpha m)^2}\}$ is a frame for $L^2(\mathbb{R})$ if and only if $\alpha\beta < 1$ (overcritical sampling)—see in this connection the celebrated series of papers by K. Seip [204, 205] and also [9, 168]. The mathematical literature on Gabor expansions and their applications has astonishingly grown in sophistication in the last forty years; we recommend the classic monograph [121] as a point of departure.

On the other hand, expansions like those in (1.1) unravel only the *discrete* facet of time-frequency analysis. Let us first elaborate more on the notion of wave packet, that is a function which does possess good localization in phase space. To be more concrete, recall that good energy concentration of a non-trivial function $g \in \mathcal{S}(\mathbb{R}^d)$ on a measurable set $X \subset \mathbb{R}^d$ is achieved if there exists $0 \le \delta_X \le 1/2$ such that

$$\left(\int_{\mathbb{R}^d \setminus X} |g(y)|^2 dy \right)^{1/2} \le \delta_X \|g\|_{L^2}.$$

The spectral content of g is well concentrated on a set $\Xi \subset \mathbb{R}^d$ if the analogous estimate is satisfied by its Fourier transform $\hat{g}$ for some small $\delta_\Xi \ge 0$. Therefore, g is concentrated on the cell (or "logon" [113]) $X \times \Xi$ in phase space and the Donoho-Stark uncertainty principle prescribes a lower bound for the measure of such cell in terms of δ_X and δ_Ξ, namely $|X||\Xi| \ge (1 - \delta_X - \delta_\Xi)^2$ [76].

The essential phase-space support of g can be moved to $(x + X) \times (\xi + \Xi)$ for any choice of $(x, \xi) \in \mathbb{R}^{2d}$ by applying a so-called *time-frequency shift* $\pi(x, \xi) = M_\xi T_x$ to g, namely as a result of the joint action of the modulation operator M_ξ and the translation operator T_x, which are respectively defined by

$$M_\xi g(y) = e^{2\pi i \xi \cdot y} g(y), \qquad T_x g(y) = g(y - x), \quad y \in \mathbb{R}^d.$$

Functions of the type $\pi(z)g$ for some fixed $z \in \mathbb{R}^{2d}$ and $g \in \mathcal{S}(\mathbb{R}^d)$ are called *Gabor wave packets* or *atoms*. We retrieve Gaussian wave packets for the choice $g(y) = e^{-\pi |y|^2}$. Note in particular that the Gabor expansion in (1.1) coincides with $\sum_{m,n \in \mathbb{Z}} c_{mn} \pi(m, n) g$.

In accordance with the program of modern harmonic analysis, the dictionary of atomic elements provided by Gabor wave packets can be used to decompose functions and operators into elementary pieces—that is the so-called *analysis* step.

The focus is then shifted to the level of phase space, where one is lead to investigate how Gabor atoms interact or how they evolve under operators. Finally, one collects all these results and tries to read the overall effect on the original objects (*synthesis*), hence coming back to the primary domain hopefully with new information.

The techniques of Gabor analysis are the backbone of all results contained in this monograph, hence in the first part—corresponding to Chaps. 3 and 4—we provide an extensive and mostly self-contained derivation of the background material.

1.1.1 The Analysis of Functions via Gabor Wave Packets

Let us briefly discuss how this program is carried out for functions and distributions. A more detailed treatment of the matter is given in Chap. 3 below.

As far as the analysis step is concerned, we can design a phase-space representation of a signal $f \in L^2(\mathbb{R}^d)$ by means of a decomposition along the uniform boxes in phase space occupied by the Gabor atoms $\pi(z)g$, $z \in \mathbb{R}^{2d}$, for some fixed $g \in L^2(\mathbb{R}^d) \setminus \{0\}$. This is the continuous analogue of the expansion (1.1) and is called the *Gabor transform* of f with *window function g*—also known as the *short-time Fourier transform* (STFT) or *sliding/windowed Fourier transform*:

$$V_g f(x, \xi) := \langle f, \pi(x, \xi)g \rangle = \int_{\mathbb{R}^d} e^{-2\pi i \xi \cdot y} f(y) \overline{g(y - x)} \, dy, \quad (x, \xi) \in \mathbb{R}^{2d},$$

$$(1.2)$$

where the brackets $\langle \cdot, \cdot \rangle$ denote the inner product in L^2—or its extension to the duality $S' - S$ in the case where f is a temperate distribution and $g \in S(\mathbb{R}^d) \setminus \{0\}$. In order to understand the heuristics behind this expression, it is worth noting that it can be equivalently recast as follows:

$$V_g f(x, \xi) = \mathcal{F}(f \cdot \overline{T_x g})(\xi) = e^{-2\pi i x \cdot \xi}(f * M_\xi g^*)(x),$$

where we set $g^*(y) = \overline{g(-y)}$. Now, fix $x \in \mathbb{R}^d$ and assume for simplicity that g is a real-valued smooth function with compact support; then $f \cdot T_x g$ is just a slice of the original signal f near the "instant" x and $V_g f(x, \xi)$ provides information on the spectral content of this piece of the signal. It is therefore clear the role of $T_x g$ as a sliding cut-off function near $x \in \mathbb{R}^d$. Due to overlaps we have that $V_g f$ is a highly redundant time-frequency representation; the design of the window function is in fact a major problem for obtaining a satisfactory resolution. For this and other problems of interest for applications we suggest the comprehensive references [21, 137].

The idea behind the STFT encodes the paradigm of local Fourier analysis, which first appeared in the signal processing community in an attempt to overcome the practical limitations of the Fourier transform—see the pioneering papers by J. Ville

[228] and the classic monographs [45, 105]. The concrete experience of listening to music is very instructive in this connection; in a sense, this just amounts to the knowledge of a signal in the time domain, where the transition between notes can be perceived but the latter cannot be identified. Conversely, thanks to the Fourier transform we can derive statistics on the abundance of single notes forming the piece, but we have no information on when (and for how long) they are in play. The short-time Fourier transform is instead a mathematical analogue of the musical score—or, in more evocative terms, a rough mathematical model of hearing.

The Gabor transform is a rich source of intriguing mathematical problems, including invertibility/reconstruction of a signal from the knowledge of its STFT and the connection with discrete samples and Gabor expansions in the spirit of (1.1) (with $c_{mn} = \langle f, \pi(m, n)g \rangle$); we refer again to [121] for a comprehensive account of these and many other issues.

It is natural at this stage to introduce new function spaces that are well-behaved under this transform in some sense. For instance, the so-called *modulation spaces* were introduced by H. G. Feichtinger in the 1980s [83, 84, 86] and can be thought of as the parallel of Besov-Triebel-Lizorkin spaces for wavelets, with uniform rather than dyadic geometry. They can be equivalently designed by constraining the summability/decay of functions in phase space, that is the global behaviour of their time-frequency representations. To be precise, for $1 \leq p \leq \infty$ the modulation space $M^p(\mathbb{R}^d)$ is the Banach space consisting of all the distributions $f \in S'(\mathbb{R}^d)$ whose Gabor transform is L^p-integrable on phase space for some (in fact, any) non-trivial window $g \in S(\mathbb{R}^d)$, that is

$$\|f\|_{M^p(\mathbb{R}^d)} := \|V_g f\|_{L^p(\mathbb{R}^{2d})} < \infty.$$

More general spaces can be defined similarly by fine-tuning of the phase-space regularity. To be precise, given a suitable *weight function* m on $\mathbb{R}^{2d}$ (basically, a continuous and positive function with at most polynomial growth), the modulation space $M_m^{p,q}(\mathbb{R}^d)$, $1 \leq p, q \leq \infty$, can be defined as above under the condition

$$\|f\|_{M_m^{p,q}} := \left(\int_{\mathbb{R}^d} \left(\int_{\mathbb{R}^d} |V_g f(x, \xi)|^p m(x, \xi)^p dx \right)^{q/p} d\xi \right)^{1/q} < \infty,$$

for some (in fact, any) non-trivial $g \in S(\mathbb{R}^d)$, with obvious modifications in the case where $p = \infty$ or $q = \infty$. We clearly recover $M^p(\mathbb{R}^d)$ in the case where $m \equiv 1$ and $p = q$. We are usually concerned with weights of polynomial type, namely for $s \in \mathbb{R}$ we set $v_s(x) := (1 + |x|)^s$, $x \in \mathbb{R}^d$. We also consider amalgamated weights of type $m(z) = v_s(z)$ or split weights like $m(x, \xi) = v_r(x)v_s(\xi)$ for $r, s \in \mathbb{R}$—in this case we write $M_{r,s}^{p,q}(\mathbb{R}^d)$ for clarity. As a rule of thumb, the decay and smoothness of $f \in M_{r,s}^{p,q}(\mathbb{R}^d)$ are related to the weighted L^p and L^q summability of $V_g f(x, \xi)$ with respect to x and ξ respectively.

These families of Banach spaces provide an optimal framework for the problems of Gabor analysis and enjoy a large number of nice properties and connections with other well-known spaces of harmonic analysis—notably $M^2(\mathbb{R}^d) = L^2(\mathbb{R}^d)$. Moreover, in the last twenty years they had a non-negligible impact on the study of pseudodifferential operators and nonlinear partial differential equations. We cannot hope to frame all the relevant literature here, we just mention the classic monographs [96, 97, 121] and the more recent ones [16, 50, 232] for a wide perspective and further references; we also recommend the survey [198] for applications to partial differential equations.

While the Gabor transform is a well-defined continuous mapping $V_g : M^p(\mathbb{R}^d) \rightarrow L^p(\mathbb{R}^{2d})$ that performs the analysis of a function in terms of Gabor wave packets, the inverse problem of synthesis/reconstruction is encoded by the so-called *adjoint Gabor transform*. Fix a non-trivial atom $\gamma \in \mathcal{S}(\mathbb{R}^d)$; for any measurable function $F : \mathbb{R}^{2d} \rightarrow \mathbb{C}$ on phase space that grows at most polynomially (i.e., $|F(z)| = O(|z|^N)$ for some positive integer N), the adjoint Gabor transform is the distribution-valued integral defined by

$$V_\gamma^* F := \int_{\mathbb{R}^{2d}} F(z)\pi(z)\gamma \, dz \in \mathcal{S}'(\mathbb{R}^d).$$

The choice of the name is justified by the identity

$$\langle V_\gamma^* F, f \rangle = \langle F, V_\gamma f \rangle, \quad f \in \mathcal{S}(\mathbb{R}^d).$$

It can be proved that the mapping V_γ^* is continuous from $L^p(\mathbb{R}^{2d})$ to $M^p(\mathbb{R}^d)$, $1 \leq p \leq \infty$, and the following crucial *inversion formula* holds if $\langle \gamma, g \rangle \neq 0$:

$$f = \frac{1}{\langle \gamma, g \rangle} V_\gamma^* V_g f = \frac{1}{\langle \gamma, g \rangle} \int_{\mathbb{R}^{2d}} V_g f(z)\pi(z)\gamma \, dz, \quad f \in M^p(\mathbb{R}^d). \tag{1.3}$$

The role of the adjoint Gabor transform in the synthesis step is thus clarified.

1.2 The Analysis of Operators via Gabor Wave Packets

The machinery developed so far can also be used to perform a wave packet analysis of operators. We give below a general outline and refer the reader to Chap. 4 for a detailed discussion.

Given a linear continuous operator $A : \mathcal{S}(\mathbb{R}^d) \rightarrow \mathcal{S}'(\mathbb{R}^d)$, if we fix two windows $g, \gamma \in \mathcal{S}(\mathbb{R}^d)$ such that $\|g\|_{L^2} = \|\gamma\|_{L^2} = 1$ and apply the inversion formula (1.3) twice we obtain

$$A = V_\gamma^* V_\gamma A V_g^* V_g = V_\gamma^* \widetilde{A} V_g,$$

where we set $\widetilde{A} = V_\gamma A V_g^*$. It turns out that $\widetilde{A}$ corresponds to a representation of the operator A on phase space; precisely, a straightforward computation shows that $\widetilde{A}$ is an integral operator satisfying

$$\widetilde{A}(V_g f)(w) = V_\gamma (Af)(w) = \int_{\mathbb{R}^{2d}} K_A(w, z) V_g f(z) dz, \quad w \in \mathbb{R}^{2d},$$

where the integral kernel is given by the so-called *Gabor matrix* of A with respect to analysis and synthesis windows g and γ respectively:

$$K_A(w, z) := \langle A\pi(z)g, \pi(w)\gamma \rangle, \quad w, z \in \mathbb{R}^{2d}.$$

Indeed, the Gabor matrix K_A can be thought of as an infinite-dimensional matrix encoding the whole information on the phase-space features of A, since it does precisely characterize how wave packets evolve and interact under the action of A.

The wave packet analysis of an operator thus amounts to a detailed investigation of the properties of the corresponding Gabor matrix. In particular, it is clear that some form of sparsity of K_A is a highly desirable property, for both theoretical and numerical purposes. Several families of operators have been examined through this lens in the last decades, including pseudodifferential operators [66, 122, 126, 189, 194, 214], Fourier integral operators [53–55, 214] and propagators associated with Cauchy problems for Schrödinger-type evolution equations [57–59, 162, 172, 214]. We also recommend the recent monograph [50] for a more systematic account.

1.2.1   The Problem of Quantization

To give a taste of the issues arising in the above discussion, let us consider a class of operators whose Gabor matrices show a peculiar behaviour, namely *Weyl pseudodifferential operators*.

The physical motivation behind the introduction of Weyl operators resides in the problem of *quantization*. In a nutshell, one is required to associate in a "robust" way a classical observable σ (i.e., a suitable function on the phase space) with a quantum observable $\mathrm{op}(\sigma)$—which is represented by a self-adjoint operator on the Hilbert space of the system in the canonical Schrödinger picture of quantum mechanics. A naive way to perform quantization is provided by the following recipe: given an observable $\sigma : \mathbb{R}^{2d} \to \mathbb{R}$, then $\mathrm{op}(\sigma)$ is obtained by formally replacing x_j with the position operator X_j and ξ_j with the momentum operator D_j:

$$X_j f(x) = x_j f(x), \quad D_j f(x) = \frac{h}{2\pi i} \frac{\partial}{\partial x_j} f(x), \quad f \in \mathcal{S}(\mathbb{R}^d), \quad j = 1, \dots, d,$$

where $h > 0$ is a small parameter (playing the character of the Planck constant). The well-posedness issues related to such a functional calculus of operators are of primary concern. It should be also emphasized that the operators X_j and D_j may be defined in a different way as long as they satisfy the canonical commutation relations

$$[X_j, X_k] = 0, \quad [D_j, D_k] = 0, \quad 2\pi[X_j, D_k] = ih\delta_{jk}I, \quad j, k = 1, \ldots, d,$$

where $[A, B] = AB - BA$ is the commutator of the operators A and B. Moreover, the correspondence $\sigma \mapsto \operatorname{op}(\sigma)$ should depend on the parameter h in such a way that the classical observable σ can be recovered by taking the "classical limit" $\lim_{h\to 0} \operatorname{op}_h(\sigma)$ in a suitable sense (this is known in physics as the *correspondence principle* [25]). For simplicity we temporarily ignore this aspect and fix $h = 1$ in this discussion, in line with the custom in harmonic analysis. Note that it is customary to introduce the reduced parameter $\hbar := h/2\pi$, hence we have $\hbar = 1/2\pi$ for the moment.

There is plenty of quantization schemes in the literature, each one with its own strengths and weaknesses. Let us commence our discussion with the case of monomial observables for concreteness, namely $\sigma(x, \xi) = x_j^m \xi_j^n$ for some $m, n \in \mathbb{N}$ and $j = 1, \ldots, d$. We immediately remark that a clear ordering problem occurs due to the non-commutativity of X_j and D_j, which is an unavoidable source of ambiguity in the definition of $\operatorname{op}(\sigma)$. In this respect, establishing a quantization rule corresponds to fixing an order for the quantization of monomials. Two options are quite natural: the *normal* and *antinormal ordering*, also called left and right (or $q - p$ and $p - q$) quantization respectively, the name being clear from the very definition:

$$x_j^m \xi_j^n \xrightarrow{\text{left}} X_j^m D_j^n, \quad x_j^m \xi_j^n \xrightarrow{\text{right}} D_j^n X_j^m.$$

A compromise between these prescriptions which favours symmetry is provided by the *Weyl quantization*:

$$\operatorname{op}_{\mathrm{w}}(x_j^m \xi_j^n) = \frac{1}{2^n} \sum_{k=0}^{n} \binom{b}{k} D_j^{n-k} X_j^n D_j^k.$$

The correspondence introduced by H. Weyl in the late 1920s [234] is usually acknowledged as the one having optimal properties. In fact, we highlight that M. de Gosson has recently made a strong case for the Born-Jordan quantization as the optimal quantization rule, cf. [60, 72]. This is an equally weighted average of the operator orderings, namely

$$\operatorname{op}_{\mathrm{BJ}}(x_j^m \xi_j^n) = \frac{1}{m+1} \sum_{k=0}^{m} X_j^{m-k} D_j^n X_j^k = \frac{1}{n+1} \sum_{k=0}^{n} D_j^{n-k} X_j^m D_j^k.$$

Note that the Weyl and Born-Jordan orderings coincide for $m = n = 1$ and both yield the operator $(X_j D_j + D_j X_j)/2$. See also the contribution by V. Turunen [227] for applications in signal analysis.

The quantization of polynomial observables ultimately reduces to the previous rules. The coverage of more general functions crucially relies on a simple though powerful remark. Let $P(X, D) = \sum_{|\alpha| \leq m} X^\alpha D_x^\alpha$ be a linear partial differential operator of order $m \in \mathbb{N}$—we use the multi-index notation here; note that $P(X, D)$ corresponds to the normal quantization of the polynomial $P(x, \xi) = \sum_{|\alpha| \leq m} x^\alpha \xi^\alpha$. The inversion formula for the Fourier transform yields

$$P(X, D)f = \int_{\mathbb{R}^{2d}} e^{2\pi i (x-y) \cdot \xi} P(x, \xi) f(y) dy d\xi, \quad f \in S(\mathbb{R}^d).$$

It is quite tempting to take this integral representation as a definition of the normal quantization of a general function σ of both position and momentum beyond the polynomial case, at least formally. The hard work of providing a rigorous and consistent framework for the study of these operators is left to the theory of *pseudodifferential operators* [17, 106, 141, 163, 181, 207, 215, 240], with the aim to relate the properties of the operator $\sigma(x, D)$ (e.g., invertibility, composition, etc.) to those enjoyed by the corresponding symbol $\sigma(x, \xi)$ at the algebraic level.

Operators arising from this generalization of the normal quantization are usually known as *Kohn-Nirenberg pseudodifferential operators*. Similarly, given a generalized phase-space *symbol* $\sigma \in S'(\mathbb{R}^{2d})$, the *Weyl quantization* prescribes that the operator $\sigma^w = \mathrm{op}_w(\sigma) \colon S(\mathbb{R}^d) \to S'(\mathbb{R}^d)$ is (formally) defined by

$$\mathrm{op}_w(\sigma) f(x) = \int_{\mathbb{R}^{2d}} e^{2\pi i (x-y) \cdot \xi} \sigma\left(\frac{x+y}{2}, \xi\right) f(y) dy d\xi. \tag{1.4}$$

To study continuity in $L^2(\mathbb{R}^d)$ and composition of operators as in (1.4) one is required to precisely adjust the regularity and decay assumptions on the symbol. Nevertheless, a straightforward computation shows that

$$\langle \mathrm{op}_w(\sigma) f, g \rangle = \langle \sigma, W(g, f) \rangle, \quad f, g \in S(\mathbb{R}^d), \tag{1.5}$$

where we have introduced the *Wigner transform* of f, g:

$$W(f, g)(x, \xi) := \int_{\mathbb{R}^d} e^{-2\pi i \xi \cdot y} f\left(x + \frac{y}{2}\right) \overline{g\left(x - \frac{y}{2}\right)} dy.$$

If $f = g$ we write Wf. Even if its appearance is not much revealing, this sort of Fourier transform of the two-point cross-correlation between f and g was mysteriously introduced by E. Wigner in a celebrated paper of 1932 [238] as a quasi-probability distribution on phase space in order to derive quantum corrections to classical statistical mechanics, where the relevant terms are functions of jointly position and momentum. It was later rediscovered in the context of signal analysis

by J. Ville [228] and eventually became a popular tool in this community because
it enjoys several properties desired from a good time-frequency representation
[151, 175]. In fact, some of such features are shared with the STFT since

$$W(f, g)(x, \xi) = 2^d e^{4\pi i x \cdot \xi} V_{g^\vee} f(2x, 2\xi),$$

where we set $g^\vee(y) := g(-y)$. Nevertheless, there is an intrinsic difference:
the Wigner transform Wf is a *quadratic* time-frequency representation—in the
sense that $W(cf) = |c|^2 f$ for any $c \in \mathbb{C}$, while the Gabor transform is linear.
Heuristically, $Wf(x, \xi)$ is interpreted as a measure of the energy content of the
signal f in a "tight" spectral band around ξ during a "short" time interval near x.
See Sect. 3.1.2 for further details.

The weak formulation of Weyl operators given in (1.5) is certainly easier to
handle than (1.4) and allows one to cover distribution symbols without further
effort. More importantly, this is also the bridge to Gabor analysis, since modulation
spaces may be used both as symbol classes as well as background where to study
boundedness of pseudodifferential operators. The basic results in this connection
may be found in [119, 121], while [50] is devoted to more advanced outcomes,
also including applications to partial differential equations. This is an extremely
convenient stage where to see the results developed in the previous section in action,
that is analysis in phase space in terms of the Gabor matrix. Indeed, it turns out that
symbols in suitable modulation spaces associate with sparsity of the Gabor matrix of
the corresponding Weyl operator. For instance (see Theorem 4.2.5 below) we have
that, for fixed $g, \gamma \in \mathcal{S}(\mathbb{R}^d)$, if $\sigma \in M_{0,s}^\infty(\mathbb{R}^{2d})$, $s \in \mathbb{R}$, then there exists a constant
$C = C(s) > 0$ such that the Gabor matrix of the corresponding Weyl operator σ^{w}
with respect to g and γ satisfies

$$|\langle \sigma^{\mathrm{w}} \pi(z)g, \pi(w)\gamma \rangle| \leq C(1 + |w - z|)^{-s}, \quad z, w \in \mathbb{R}^{2d}.$$

Roughly speaking, we could read this result by saying that Gabor wave packets
approximately diagonalize pseudodifferential operators with symbols in $M_{0,s}^\infty(\mathbb{R}^{2d})$.
This sparsity estimate can be used to prove several types of results; for instance, it
is not difficult to show that σ^{w} is continuous on any modulation space $M^p(\mathbb{R}^d)$
provided that $\sigma \in M_{0,s}^\infty(\mathbb{R}^{2d})$ and $s > 2d$. Recall indeed that lifting the continuity
problem to the level of phase space by means of the Gabor transform amounts
to study the continuity of a phase-space representation of σ^{w}, that is an integral
operator with integral kernel given by the Gabor matrix $K_{\sigma^{\mathrm{w}}}$. Using the estimate
above, we see that the action of σ^{w} on phase space is essentially that of a convolution
operator with kernel v_{-s}, that is continuous on $L^p(\mathbb{R}^{2d})$, $1 \leq p \leq \infty$, provided that
the kernel is integrable, namely $s > 2d$. We refer to Theorem 4.2.4 for further details
and related results. Nevertheless, we believe that this easy and intuitive argument
fully witnesses the spirit of the techniques used in this field.

A similar quasi-diagonalization result holds for symbols that belong to the
modulation space $M^{\infty,1}(\mathbb{R}^{2d})$, traditionally called the *Sjöstrand class*. This space
of rough symbols (essentially, bounded continuous functions which locally coincide

with the Fourier transform of an integrable function) was discovered by J. Sjöstrand in 1994 [209] in an attempt to extend the well-behaved Hörmander class $S_{0,0}^0$. He showed that this exotic symbol class still yield pseudodifferential operators which are bounded on $L^2(\mathbb{R}^d)$. A subsequent time-frequency analysis of the Sjöstrand class [122] by K. Gröchenig showed that $\sigma \in M^{\infty,1}(\mathbb{R}^{2d})$ implies the existence of a controlling function $H \in L^1(\mathbb{R}^{2d})$ for the Gabor matrix of σ^w, that is

$$|\langle \sigma^w \pi(z)g, \pi(w)\gamma \rangle| \le H(w - z), \quad w, z \in \mathbb{R}^{2d}.$$

Note that the proof of the boundedness on any modulation space M^p mentioned before applies unchanged also in this case, hence recovering and extending Sjöstrand's result on the boundedness of σ^w in $L^2(\mathbb{R}^d) = M^2(\mathbb{R}^d)$.

1.2.2 Metaplectic Operators

Another class of operators that has been largely analyzed from a phase-space perspective is that of *metaplectic operators*. This is indeed a traditional problem of harmonic analysis in phase space [71, 106] that has deep connections with several other topics, ranging from the abstract theory of covering spaces and group representations to the study of more concrete integral operators.

There are at least three equivalent ways to introduce metaplectic operators: in terms of the Schrödinger representation of the Heisenberg group [106], by means of oscillatory integral operators (the so-called quadratic Fourier transforms [71]) or in terms of evolution operators for the Schrödinger equation with quadratic Hamiltonian [193]. Roughly speaking, the (projective) metaplectic representation is a recipe to associate a symplectic matrix S with a unitary operator $\mu(S)$ on $L^2(\mathbb{R}^d)$, unique up to a phase factor, in a product-preserving way (up to a phase factor). A rigorous and complete coverage of this topic is undoubtedly a demanding technical exercise. We now give an informal outline inspired by the third approach, while in Sect. 4.3 we discuss the issue in more detail. We wish to emphasize that our approach allows one to develop the theory from the ground up, resulting in a completely self-contained treatment that seems to be suitable for pedagogical purposes.

As mentioned, let us focus on the Schrödinger equation with quadratic Hamiltonian—cf. Sect. 4.3.3 for further details. Let Q be a real quadratic form on $\mathbb{R}^{2d}$, namely

$$Q(x, \xi) = \frac{1}{2}A\xi \cdot \xi + Bx \cdot \xi + \frac{1}{2}Cx \cdot x,$$

where $A, B, C \in \mathbb{R}^{d \times d}$ and A, C are symmetric matrices. It is not difficult to compute the corresponding Weyl quantization, that is

$$Q^{\mathrm{w}} = -\frac{1}{8\pi^2} \sum_{j,k=1}^{d} A_{j,k} \partial_j \partial_k - \frac{i}{2\pi} \sum_{j,k=1}^{d} B_{j,k} x_j \partial_k + \frac{1}{2} \sum_{j,k=1}^{d} C_{j,k} x_j x_k - \frac{i}{4\pi} \mathrm{Tr}(B).$$

Consider now the Cauchy problem in $\mathcal{S}(\mathbb{R}^d)$ for the Schrödinger equation with Hamiltonian given by Q^{w} (and the normalization $\hbar = 1/2\pi$), namely

$$\begin{cases} \frac{i}{2\pi} \partial_t \psi = Q^{\mathrm{w}} \psi \\ \psi(0, x) = f(x), \end{cases} \qquad (t, x) \in \mathbb{R} \times \mathbb{R}^d.$$

It turns out that the corresponding unitary propagator, that is the evolution operator $U(t) = e^{-2\pi i t Q^{\mathrm{w}}}$, is a metaplectic operator—in fact, all the metaplectic operators can be characterized as the composition of such operators. Precisely, $U(t)$ is the metaplectic operator associated with the symplectic matrix S_t arising as the Hamiltonian flow $\mathbb{R} \ni t \mapsto S_t \in \mathrm{Sp}(d, \mathbb{R})$ corresponding to the classical system with Hamiltonian Q. Figuratively speaking, the metaplectic representation is a correspondence between the classical scenario and the corresponding quantum counterpart; we refer to Theorem 4.3.9 for a rigorous formulation.

The Gabor analysis of a metaplectic operator $\mu(S)$ is particularly convenient in view of the following sparsity estimate for the corresponding Gabor matrix (cf. Theorem 4.3.13): for fixed windows $g, \gamma \in \mathcal{S}(\mathbb{R}^d)$ and any $N \geq 0$ we have

$$|\langle \mu(S) \pi(z) g, \pi(w) \gamma \rangle| \leq C(1 + |w - Sz|)^{-N}, \quad w, z \in \mathbb{R}^{2d},$$

for some constant $C > 0$ that depends on N and S. Therefore, the entries of the Gabor matrix of $\mu(S)$ show a rapid decay away from the graph in phase space of the corresponding matrix S; note that for the Schrödinger propagator $U(t)$ discussed before, the last estimate shows that the quantum evolution in phase space approximately follows the classical one corresponding to S_t, again in accordance with Bohr's correspondence principle.

In general, a linear continuous operator $\mathcal{S}(\mathbb{R}^d) \to \mathcal{S}'(\mathbb{R}^d)$ whose Gabor matrix satisfies estimates of this type will be called *generalized metaplectic operator*. Such operators can be characterized in more classical terms as suitable oscillatory integral operators or Fourier integral operators; they will play a key role in the second part of the monograph. In the same spirit, even more general classes of operators $FIO(\chi, v_s)$ are introduced below, consisting of operators characterized by a similar estimate for the Gabor matrix, with N replaced by a given $s \geq 0$ and S by a suitable, possibly non linear, symplectomorphism $\chi : \mathbb{R}^{2d} \to \mathbb{R}^{2d}$.

To conclude this short summary of the fundamental techniques of Gabor analysis, we wish to say a few words on *semiclassical* time-frequency analysis. We already mentioned before, when discussing the problem of quantization, the role of the asymptotics $h \to 0$ for the Planck parameter. More generally, a quite inaccurate yet effective way to state Bohr's correspondence principle [25] is as follows: the predictions of quantum mechanics should agree with those of classical physics in the limit $h \to 0$. The occurrence of small parameters and the interest in their asymptotic behaviour is also a common issue in the theory of PDEs—cf. again the Planck constant h in the Schrödinger equation.

In order to make the tools of Gabor analysis suitable to deal with these problems, it is necessary to reconsider all the operators and function spaces introduced so far and to carefully keep track of the Planck parameter h (or the so-called reduced version $\hbar = h/2\pi$). This program is carried out in Chap. 5, at least for the needs of this monograph.

1.3   The Problem of Feynman Path Integrals

There has been plenty of opportunities to discuss the fruitful interplay between Gabor analysis and physics for what concerns motivations as well as applications; the problem of quantization is perhaps the most striking example in this respect. In fact, another problem arising in mathematical physics has recently benefited from the techniques of Gabor analysis, that is the rigorous formulation of *Feynman path integrals*. From the mathematical point of view, while the quantization problem deals with the characterization of an operator A in terms of its symbol a and the correspondence $a \mapsto A$, path integrals are basically a way to provide explicit representation formulas for the evolution operator e^{-itA} (and also e^{-tA} in general).

The path integral formulation of non-relativistic quantum mechanics is a cornerstone of modern theoretical physics, due to R. Feynman (Nobel Prize in Physics 1965). The first appearance of this approach dates back to a 1948 paper [102], but its origins are to be found in Feynman's Ph.D. thesis (Princeton University, 1942)—cf. the recent reprint [30] and [199] for additional historical details. It would be a shame not to give at least a mention to Feynman diagrams, introduced in a celebrated subsequent article [103], which have definitely reshaped the whole quantum field theory. A masterly introduction to this beautiful circle of ideas can be found in a nowadays classical textbook, authored by Feynman himself and his former graduate student A. Hibbs [104]. We anticipate that the motivation to the introduction of path integrals comes from an insightful, far-reaching elaboration on the two-slit experiment.

Let us briefly describe the main features of Feynman's approach. Broadly speaking, one could argue that while the standard quantization framework (as developed by Dirac) relies on the Hamiltonian formulation of classical mechanics, path integrals largely draw on the ideas of Lagrangian mechanics and ultimately provide a sort of quantum counterpart of the latter. Recall that the state of a non-relativistic particle in the Euclidean space $\mathbb{R}^d$ at time t is represented by the wave function $\psi(t, x)$, $(t, x) \in \mathbb{R} \times \mathbb{R}^d$, such that $\psi(t, \cdot) \in L^2(\mathbb{R}^d)$. The time evolution of a state $f(x)$ at $t = 0$ under a real-valued potential V is governed by the Cauchy problem for the Schrödinger equation:

$$\begin{cases} i\hbar\partial_t\psi = (H_0 + V(t,x))\psi \\ \psi(0,x) = f(x), \end{cases} \tag{1.6}$$

where $H_0 = -\hbar^2\Delta/2$ is the free particle Hamiltonian. Note that we conveniently set $m = 1$ for the mass of the particle. The corresponding Schrödinger propagator[1] $U(t,s) : \psi(s,\cdot) \mapsto \psi(t,\cdot)$, $t, s \in \mathbb{R}$, is a unitary operator on $L^2(\mathbb{R}^d)$; we set $U(t)$ for $U(t,0)$. Since $U(t)$ is a linear operator, at least on a formal level we can represent it as an integral operator:

$$\psi(t,x) = \int_{\mathbb{R}^d} u_t(x,y) f(y) dy,$$

where the kernel $u_t(x,y)$ represents the transition amplitude from the position y at time 0 to the position x at time t. To put it simply, path integrals show up in a pretty natural way if one tries to take into account all the possible "alternative histories" of the system, that is to say the infinite number of trajectories γ with fixed endpoints ($\gamma(0) = y$ and $\gamma(t) = x$) that the particle could follow. According to Feynman's prescription, the contribution to the transition amplitude of each interfering path is a phase factor that involves the *action functional* along the path (in units of $\hbar$), that is

$$S[\gamma] = \int_0^t L(\gamma(\tau), \dot{\gamma}(\tau), \tau) d\tau, \tag{1.7}$$

where $L(x,v,t) = |v|^2/2 - V(x,t)$ is the Lagrangian of the underlying classical system. As a result, the kernel has the following representation:

$$u_t(x,y) = \int e^{\frac{i}{\hbar}S[\gamma]} \mathcal{D}\gamma, \tag{1.8}$$

a formal expression entailing a sort of integral over the infinite-dimensional manifold of paths with endpoint constraints as detailed above. This suggestive, yet unorthodox, picture is surprisingly reinforced by the following remark: glossing over mathematical rigour for another moment, a formal argument relying on the stationary phase principle shows that the semiclassical limit $\hbar \to 0$ actually selects the classical trajectory, consistently with the principle of stationary action of classical mechanics (Fig. 1.1).

---

[1] A word of caution on conventions is needed here, as we are in presence of a possible source of confusion. In the physics literature the term "propagator" is usually reserved to the integral kernel u_t of $U(t)$, see below. This clearly conflicts with the custom in the analysis of PDEs, where the propagator denotes in fact $U(t,s)$. We join the latter tradition in this book.

Fig. 1.1 Some of all the possible trajectories with given endpoints $(0, x_0)$ and (t, x) that a particle could follow (in red). In blue: the classical path, which is recovered in the semiclassical limit $\hbar \to 0$

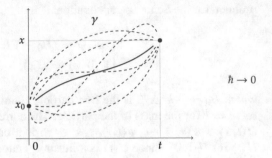

While we used the familiar integral symbol to condense Feynman's arguments, it was proved by R. Cameron [34] that $\mathcal{D}\gamma$ cannot be a Lebesgue-type measure on the space of paths, neither one can arrange a Wiener measure with complex variance for the same purpose as it would have infinite total variation [174]. Path integrals are still today a source of challenges to the mathematical community, in a joint effort to put Feynman's compelling intuition on firm mathematical ground. In particular, the literature on the attempts to interpret (1.8) as an *infinite-dimensional oscillatory integral* is huge; the interested reader could certainly benefit from the monographs [5, 174] as points of departure.

Here we prefer to focus on the original *time slicing* heuristics [104], namely we consider a subdivision of the time interval and then compute the action functional on each sub-interval along suitably chosen classes of paths. We will describe below two of the several approaches that rely on this principle, which happen to be two faces of the same operator-algebraic concept that can be stated as follows. One is ultimately lead to design sequences of approximate propagators on $L^2(\mathbb{R}^d)$ that are well behaved under composition and sufficient short-time approximation power. Such sequences are expected to converge to the exact propagator $U(t)$ in some sense (i.e., topology) to be specified, the latter competing against the regularity properties of the potential V.

We can perceive in this formulation the flavor of a problem of classical harmonic analysis. This inspiration comes out both with respect to the type of problems that we are going to consider (e.g., pointwise convergence of the integral kernels or convergence in the strong/uniform operator topology) as well as the techniques used for their solution. Indeed, this is exactly the point where Gabor analysis enters the scene: the accumulated knowledge on the wave packet analysis of operators will be crucial in the study of approximate propagators arising in the theory of path integrals. Moreover, modulation spaces enjoy a satisfactory balance between nice properties (such as the Banach algebra structures, decomposition, embeddings, etc.) and regularity of their members, so that they can be used as reservoirs of potentials.

We now elaborate on the problems that are addressed in the second part of this monograph.

1.3.1   Rigorous Time-Slicing Approximation of Feynman Path Integrals

The closest approach to Feynman's original intuition (cf. [104]) requires interpreting formula (1.8) by means of a limiting procedure involving suitably designed finite-dimensional approximations.

To be more precise, let us briefly recall the elements of the sequential approach (cf. [174]) in the form introduced by E. Nelson in [178], which is based upon two basic results. First, recall that the Schrödinger propagator $U_0(t) = e^{-\frac{i}{\hbar}t H_0}$ for the free particle Hamiltonian $H_0 = -\hbar^2 \Delta /2$, is a Fourier multiplier; routine calculations lead to the following integral representation [191, Section IX.7]:

$$e^{-\frac{i}{\hbar}t H_0} f(x) = \frac{1}{(2\pi i t \hbar)^{d/2}} \int_{\mathbb{R}^d} \exp\left(\frac{i}{\hbar} \frac{|x-y|^2}{2t}\right) f(y) dy, \qquad f \in \mathcal{S}(\mathbb{R}^d).$$

(1.9)

It is worth emphasizing here a vague resemblance with (1.8), further enhanced by an apparently mysterious occurrence. Note indeed that the phase factor in the integral equals the action functional evaluated along the line $\gamma_{cl}(\tau) = y + (x - y)\tau/t$, the latter being precisely the classical trajectory of a particle moving from position y at time $\tau = 0$ to position x at time $\tau = t$ in the absence of external forces.

Suppose that the potential $V(x)$ in time-independent. We then resort to a result from the theory of operator semigroups, namely the *Trotter product formula* holds for the propagator $U(t) = e^{-\frac{i}{\hbar}t(H_0+V)}$ generated by $H = H_0 + V$:

$$U(t)f = \lim_{n\to\infty} E_n(t)f, \quad f \in L^2(\mathbb{R}^d), \qquad E_n(t) := \left(e^{-\frac{i}{\hbar}\frac{t}{n}H_0} e^{-\frac{i}{\hbar}\frac{t}{n}V}\right)^n,$$

(1.10)

which holds under suitable conditions on the domain of H_0 and on the potential V (see below). We have thus convergence of the approximate propagators $E_n(t)$ (also called *Feynman-Trotter parametrices*) in the *strong topology* of operators in $L^2(\mathbb{R}^d)$ to the exact propagator $U(t)$. Combining such ingredients yields the following representation of the full propagator $e^{-\frac{i}{\hbar}t H}$ as limit of a sequence of integral operators [191, Theorem X.66]:

$$e^{-\frac{i}{\hbar}t(H_0+V)} f(x) = \lim_{n\to\infty} \left(2\pi\hbar i \frac{t}{n}\right)^{-\frac{nd}{2}} \int_{\mathbb{R}^{nd}} e^{\frac{i}{\hbar} S_n(t;x_0,\dots,x_{n-1},x)} f(x_0) dx_0 \dots dx_{n-1},$$

(1.11)

where we set

$$S_n(t; x_0, \ldots, x_{n-1}, x) = \sum_{k=1}^{n} \frac{t}{n} \left[\frac{1}{2} \left(\frac{|x_k - x_{k-1}|}{t/n} \right)^2 - V(x_k) \right], \quad x_n = x.$$

In order to elucidate the role of $S_n(t; x_0, \ldots, x_n)$ we argue as follows. Given the points $x_0, \ldots, x_{n-1}, x \in \mathbb{R}^d$, let $\overline{\gamma}$ be the broken line (i.e., polygonal path) with vertices $x_k = \overline{\gamma}(kt/n), k = 0, \ldots, n-1, x_n = x$, with the parametrization

$$\overline{\gamma}(\tau) = x_k + \frac{x_{k+1} - x_k}{t/n} \left(\tau - k\frac{t}{n} \right), \quad \tau \in \left[k\frac{t}{n}, (k+1)\frac{t}{n} \right], \quad k = 0, \ldots, n-1.$$
$$(1.12)$$

Therefore, $\overline{\gamma}$ entails a motion with constant velocity along each segment (Fig. 1.2). The action along such path is

$$S[\overline{\gamma}] = \sum_{k=1}^{n} \frac{1}{2} \frac{t}{n} \left(\frac{|x_k - x_{k-1}|}{t/n} \right)^2 - \int_0^t V(\overline{\gamma}(\tau)) d\tau.$$

In accordance with Feynman's heuristics, formula (1.11) can be conveniently interpreted as a way to define a path integral on the space of polygonal paths, where $S_n(x_0, \ldots, x_n, t)$ is an approximation of the underlying action functional and the integral of $V(\overline{\gamma}(\tau))$ is replaced by a corresponding Riemann-like discretization. The limiting behaviour for $n \to \infty$ then appears consistent with (1.8), at least intuitively, as any possible path is approximated by arbitrarily tight-fitting broken lines. As a matter of fact, the custom in the physics community after Feynman is exactly to think of (1.8) as a convenient notation for the broken-line approximation underlying (1.11) and the related arguments—see for instance [128, 158].

Let us briefly comment on the assumptions on the potential V under which the Trotter product formula holds. A standard sufficient condition is that V is chosen in such a way that $H_0 + V$ is essentially self-adjoint on $D = D(H_0) \cap D(V)$ in $L^2(\mathbb{R}^d)$, cf. for instance [190, Theorem VIII.31]. The powerful sequential approach allows one to cover notable classes of wild perturbations, such as the so-called *Kato potentials*, including finite sums of real-valued functions in $L^p(\mathbb{R}^d)$ with $2p > d$ and $p \geq 2$ [178, Theorem 8]. A simpler variant that best fits our purposes

Fig. 1.2 The broken line approximation $\overline{\gamma}$ introduced in (1.12)

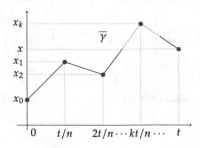

requires H_0 to be a self-adjoint operator on the domain $D(H_0) \subset L^2(\mathbb{R}^d)$, while $V \in \mathcal{L}(L^2(\mathbb{R}^d))$ is a bounded perturbation, cf. Theorem 6.4.1.

A different choice of the class of admissible paths leads to a different form of time-slicing approximation that could be informally referred to as "the Japanese way" to rigorous path integrals, in honour of the pioneers in its development: D. Fujiwara and N. Kumano-go, with further contributions by W. Ichinose, T. Tsuchida and K. Yajima. The main references for this approach are the papers [108, 109, 112, 143, 144, 164, 165, 226, 241] and the monograph [111], to which the reader is addressed for further details.

Let us focus again on Eq. (1.11) and its interpretation of a path integral in terms of finite-dimensional approximation along broken lines. We will show in a moment that a result in the same spirit can be obtained without resorting to the Trotter formula. Let us anticipate the class of (possibly time-dependent) potentials that we are allowed to consider.

Assumption (A) *The potential* $V : \mathbb{R} \times \mathbb{R}^d \to \mathbb{R}$ *satisfies* $\partial_x^\alpha V \in C^0(\mathbb{R} \times \mathbb{R}^d)$ *for any* $\alpha \in \mathbb{N}^d$ *and*

$$|\partial_x^\alpha V(t, x)| \leq C_\alpha, \quad |\alpha| \geq 2, \quad (t, x) \in \mathbb{R} \times \mathbb{R}^d,$$

for suitable constants $C_\alpha > 0$.

In particular, $V(t, x)$ is smooth in x and has at most quadratic growth. Consider then the Hamiltonian

$$H(t, x, \xi) = \frac{1}{2}|\xi|^2 + V(t, x).$$

We denote by $(x(t, s, y, \eta), \xi(t, s, y, \eta))$ $(s, t \in \mathbb{R}, y, \eta \in \mathbb{R}^d)$, the solution of the corresponding classical equations of motion

$$\begin{cases} \dot{x} = \xi \\ \dot{\xi} = -\nabla_x V(t, x) \end{cases}$$

with initial condition at time $t = s$ given by $x(s, s, y, \eta) = y, \xi(s, s, y, \eta) = \eta$. The flow

$$(x(t, s, y, \eta), \xi(t, s, y, \eta)) = \chi(t, s)(y, \eta) \tag{1.13}$$

defines a smooth canonical transformation $\chi(t, s) \colon \mathbb{R}^{2d} \to \mathbb{R}^{2d}$ satisfying for every $T_0 > 0$ the estimates

$$|\partial_y^\alpha \partial_\eta^\beta x(t, s, y, \eta)| + |\partial_y^\alpha \partial_\eta^\beta \xi(t, s, y, \eta)| \leq C_{\alpha,\beta}(T_0), \quad y, \eta \in \mathbb{R}^d, \tag{1.14}$$

for some constant $C_{\alpha,\beta}(T_0) > 0$, if $|t - s| \le T_0$ and $|\alpha| + |\beta| \ge 1$ (see [109, Proposition 1.1]). In particular, the flow is globally Lipschitz and the same holds for its inverse.

Moreover, there exists $\delta > 0$ such that for $0 < |t - s| \le \delta$ and every $x, y \in \mathbb{R}^d$, there exists only one classical path γ such that $\gamma(s) = y$, $\gamma(t) = x$. By computing the action functional along this path γ, as in (1.7), we define the generating function

$$S(t, s, x, y) = S[\gamma] = \int_s^t L(\gamma(\tau), \dot{\gamma}(\tau), \tau)d\tau, \tag{1.15}$$

for $0 < |t - s| \le \delta$.

Then Fujiwara showed [109] that the propagator $U(t, s)$ is an *oscillatory integral operator* (for short, OIO) provided that $|t - s|$ is small enough, that is

$$U(t, s)f(x) = \frac{1}{(2\pi i\hbar(t - s))^{d/2}} \int_{\mathbb{R}^d} e^{\frac{i}{\hbar}S(t,s,x,y)} a(\hbar, t, s)(x, y)f(y)dy, \tag{1.16}$$

for an amplitude function $a(\hbar, t, s) \in C_b^\infty(\mathbb{R}^{2d})$—the space of smooth functions with bounded derivatives of any order, also known as the Hörmander class $S_{0,0}^0$ in microlocal analysis. Moreover, $a(\hbar, t, s)$ is such that $\partial_x^\alpha \partial_y^\beta a(\hbar, t, s)(x, y)$ is of class C^1 in t, s, with

$$\|a(\hbar, t, s)\|_{C_b^m} := \sup_{|\alpha| \le m} \|\partial^\alpha a(\hbar, t, s)\|_{L^\infty} \le C_m,$$

for $0 < t - s \le \delta, 0 < \hbar \le 1, m \in \mathbb{N}$.

In concrete situations, except for a few simple cases, there is no feasible way to obtain a manageable explicit formula for the exact propagator. In view of this difficulty and motivated by the case of the free particle (1.9), it is natural to consider approximate propagators (*parametrices*) having the form

$$E^{(0)}(t, s)f(x) = \frac{1}{(2\pi i\hbar(t - s))^{d/2}} \int_{\mathbb{R}^d} e^{\frac{i}{\hbar}S(t,s,x,y)} f(y)dy. \tag{1.17}$$

This operator is supposed to provide a good approximation of $U(t, s)$ at least for $|t - s|$ small enough. The case of a large interval $|t - s|$ can be treated accordingly via a time slicing approach, namely by means of composition of such approximate propagators on sub-intervals. To be precise, given a subdivision $\Omega = \{t_0, \ldots, t_L\}$ of the interval $[s, t]$ such that $s = t_0 < t_1 < \ldots < t_L = t$ and $t_j - t_{j-1} \le \delta$, we define the operator

$$E^{(0)}(\Omega, t, s) = E^{(0)}(t_L, t_{L-1})E^{(0)}(t_{L-1}, t_{L-2}) \cdots E^{(0)}(t_1, t_0),$$

whose integral kernel $e^{(0)}(\Omega, t, s)(x, y)$ can be explicitly computed from (1.17), namely

$$e^{(0)}(\Omega, t, s)(x, y)$$

$$= \prod_{j=1}^{L} \frac{1}{(2\pi i(t_j - t_{j-1})\hbar)^{d/2}} \int_{\mathbb{R}^{d(L-1)}} \exp\left(\frac{i}{\hbar} \sum_{j=1}^{L} S(t_j, t_{j-1}, x_j, x_{j-1})\right) \prod_{j=1}^{L-1} dx_j,$$

with $x = x_L$ and $y = x_0$. A detailed analysis can be found in [111, Chapter 2].

The parametrix $E^{(0)}(\Omega, t, s)$ is therefore expected to converge, in a sense to be specified, to the actual propagator $U(t, s)$ in the vanishing mesh size limit, that is when

$$\omega(\Omega) := \max\{t_j - t_{j-1}, \ j = 1, \dots, L\} \to 0.$$

Note that this scheme is definitely more sophisticate than Nelson's one discussed before, also because the broken lines are now replaced by refined piecewise classical paths (Fig. 1.3). This fact will be relevant in view of the semiclassical limit, as we will show in a moment. However, it is worth mentioning that a quite complete theory of path integration for approximations on straight lines in this spirit has been developed by Kumano-go [165].

Let us now outline two milestone results from the pioneering papers by Fujiwara. In [108] convergence of $E^{(0)}(\Omega, t, s)$ to $U(t, s)$ in the norm operator topology in $\mathcal{L}(L^2(\mathbb{R}^d))$ was proved. Remarkably, under the same assumptions, a (very strong) form of convergence at the level of integral kernels was achieved in [109]. In fact, it should be emphasized that similar results are obtained for the higher-order parametrices $E^{(N)}(t, s)$, $N = 1, 2, \dots$, also known as *Birkhoff-Maslov parametrices* [19, 173] and defined by

$$E^{(N)}(t, s)f(x) = \frac{1}{(2\pi i\hbar(t - s))^{d/2}} \int_{\mathbb{R}^d} e^{\frac{i}{\hbar}S(t, s, x, y)} a^{(N)}(\hbar, t, s)(x, y)f(y)dy,$$

$$(1.18)$$

Fig. 1.3 A piecewise classical path in spacetime

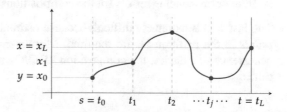

where $a^{(N)}(\hbar, t, s)(x, y) = \sum_{j=1}^{N} (\frac{i}{\hbar})^{1-j} a_j(t, s)(x, y)$ for suitable functions $a_j(t, s) \in C_b^{\infty}(\mathbb{R}^{2d})$ uniformly with respect to $0 < \hbar \leq 1, 0 < |t - s| \leq \delta$.

The case of a long interval $|t - s|$ can be treated as before by means of composition over a sufficiently fine subdivision $\Omega = \{t_0, \ldots, t_L\}$ of the interval $[s, t]$ such that $s = t_0 < t_1 < \ldots < t_L = t$, namely

$$E^{(N)}(\Omega, t, s) = E^{(N)}(t_L, t_{L-1}) E^{(N)}(t_{L-1}, t_{L-2}) \cdots E^{(N)}(t_1, t_0). \tag{1.19}$$

The L^2 theory for the time-slicing approximation can be condensed as follows.

Theorem 1.3.1 *Let the potential V satisfy Assumption (A) and fix $T_0 > 0$. For $0 < t - s \leq T_0$ and any subdivision Ω of the interval $[s, t]$ such that $\omega(\Omega) \leq \delta$, the following claims hold.*

(i) There exists a constant $C = C(N, T_0) > 0$ such that

$$\left\| E^{(N)}(\Omega, t, s) - U(t, s) \right\|_{L^2 \to L^2} \leq C \hbar^N \omega(\Omega)^{N+1} (t - s), \quad N \in \mathbb{N}.$$
$$\tag{1.20}$$

(ii) There exists $C = C(m, N, T_0) > 0$ such that

$$\left\| a(\hbar, t, s) - a^{(N)}(\Omega, \hbar, t, s) \right\|_{C_b^m} \leq C \hbar^N \omega(\Omega)^{N+1} (t - s), \quad m, N \in \mathbb{N},$$

cf. (1.16). In particular,

$$\lim_{\omega(\Omega) \to 0} a^{(N)}(\Omega, \hbar, t, s) = a(\hbar, t, s) \quad \text{in } C_b^{\infty}(\mathbb{R}^{2d}).$$

As already anticipated, the proof of these results relies on a delicate analysis of OIOs. In short, the main steps are:

1. Prove that the "time slicing approximation is an oscillatory integral" (cf. [111]), namely that the operators in (1.18) are well-defined OIOs under suitable assumptions.
2. Derive precise results and estimates for such OIOs.
3. Infer corresponding results for the compositions in (1.19).

The last step is extremely difficult because estimates uniform in L, the number of points in the partition Ω, are required. Moreover, due to phenomena such as the occurrence of caustics, the composition of OIOs results in an OIO only for short times.

1.3.2   Pointwise Convergence at the Level of Integral Kernels for Feynman-Trotter Parametrices

The sequential approach is a powerful way to give a rigorous interpretation of Feynman path integrals, as explained so far. Nevertheless, there are still reasons for not being completely satisfied with the current situation. Above all, a closer look at Feynman's original paper [102] and the textbook [104] indicates that the time slicing approximation procedure is carried out at the level of integral kernels rather than operators. Equivalently, one should primarily aim at *pointwise* convergence of the integral kernels of the approximate propagators to the kernel u_t of the Schrödinger propagator, rather than convergence in some operator topology. This crucial remark compels us to shift the focus of the previous discussions from the operators to their kernels. In doing so, we can formulate a clear and ambitious research agenda: in the framework of Nelson's approach this means that the approximation operators $E_n(t)$ should be explicitly characterized as integral operators first, at least in the sense of distributions in view of the Schwartz kernel theorem, then one should determine whether the corresponding kernels $e_{n,t}$ are genuine functions and eventually discuss the convergence to the kernel u_t of the exact propagator $U(t)$.

This program is carried out in Chap. 6, that is a refined and expanded version of the analysis in [55, 99, 183]. As already observed, Theorem 1.3.1 above represents a significant achievement on the convergence at the level of integral kernels. However, although highly sophisticated, the potential is required to be smooth and thus the result does not apply to the most popular class of potentials considered in the path integral literature, cf. [5, 146, 174], namely that of functions whose Fourier transform is a complex (finite) measure. Instead, we will see that the fruitful exploitation of the techniques of Gabor analysis discussed in the previous sections allows one to obtain a pointwise convergence result at the level of integral kernels for such potentials.

Specifically, we consider the problem (1.6) and the corresponding Trotter formula (1.10) in the setting where $H_0 = Q^{w,\hbar}$ is the $\hbar$-Weyl quantization— the semiclassical version of Weyl quantization—of a real-valued quadratic form Q; hence a possible linear magnetic potential (uniform magnetic field) and a quadratic electric potential are included in H_0 and the operator $e^{-\frac{i}{\hbar}tH_0}$ turns out to be a metaplectic operator $\mu^\hbar(S_t)$, where $\mu^\hbar$ is the semiclassical version of the metaplectic representation μ in Sect. 1.2.2 and S_t is the phase-space flow associated with the classical Hamiltonian Q. The low-regular potential $V(x)$ is assumed to be independent of time and belongs to any of the modulation spaces $M_{0,s}^\infty(\mathbb{R}^d)$, $s > 2d$, or $M^{\infty,1}(\mathbb{R}^d)$—or, more generally, V is a non-local potential in the form of a $\hbar$-Weyl operator with symbol in $M_{0,s}^\infty(\mathbb{R}^{2d})$, $s > 2d$, or $M^{\infty,1}(\mathbb{R}^{2d})$. Recall from Sect. 1.2.1 that pseudodifferential operators in such spaces are particularly well behaved when viewed from a phase-space perspective. Indeed, the success of the wave packet analysis is definitely due to the fact that Gabor wave packets approximately diagonalize both $e^{-\frac{i}{\hbar}\frac{t}{n}H_0}$ and $e^{-\frac{i}{\hbar}\frac{t}{n}V}$.

As far as the schedule above is concerned, first we use results of pseudodifferential calculus and a number of technical lemmas to prove that the Feynman-Trotter parametrix $E_n(t)$ in (1.10) can be ultimately recast as the composition of a $\hbar$-Weyl operator with symbol in the aforementioned classes and the operator $e^{-\frac{i}{\hbar}tH_0}$. Now, up to a set of exceptional times, we provide an explicit representation of $E_n(t)$ as an oscillatory integral operator and then we derive a manageable expression for its distribution kernel $e_{n,t}$. In fact, we prove that convergence of $e_{n,t}$ occurs in a certain strong sense, depending on the regularity level set by the choice of the potential; in any case, it turns out that uniform convergence on compact subsets, hence pointwise convergence too, do hold at the level of integral kernels. We observe that the operators $E_n(t)$, as well as $U(t)$, fall precisely in the class of generalized metaplectic operators mentioned before, which gives the relevant integral representation and information on the structure of the so-called fundamental solution. Notice that standard tools from classical harmonic analysis, such as stationary phase techniques, are not available at this low level of regularity.

We highlight that a convergence result in the same spirit will be also derived for smooth potentials with bounded derivatives of any order; this should be compared with Theorem 1.3.1, where the proof is certainly more sophisticate. Finally, we also partially extend the previous results to exceptional times, namely those for which the kernel of the propagator $U(t)$ is a genuine distribution and thus the problem of convergence at the level of kernels is pointless; still, we provide convergence results in a suitable modulation space norm which are better than rather general convergence in the sense of distributions.

1.3.3   Convergence of Time-Slicing Approximations in $\mathcal{L}(L^2)$ for Low-Regular Potentials

The results by Fujiwara that we have briefly collected above are certainly a paramount contribution to the mathematical theory of Feynman path integrals. Nevertheless, there are some issues that deserve further investigation. A moment of thought on the content of Theorem 1.3.1, in particular on the occurrence of convergence results at two different levels—a coarser one (parametrices in $\mathcal{L}(L^2(\mathbb{R}^d))$) and a finer one (OIO amplitudes in $C_b^\infty(\mathbb{R}^{2d})$)—suggests that the assumption on the regularity of the potential may be relaxed to some extent while still retaining convergence of the time-slicing approximations in the norm operator topology.

This is precisely the *leitmotiv* of the results presented in Chaps. 7 and 8. In spite of the common motivation, they differ both for the focus and the techniques used. It may thus be convenient to provide a short summary where to highlight the key points.

In Chap. 7, which is based on the papers [74, 169–171, 182], the potential $V(t, x)$ is required to be a bounded continuous function together with a certain number of derivatives in the Sjöstrand class $M^{\infty,1}$.

In that chapter we address the problem of bridging the gap between the delicate construction of the time-slicing approximations discussed before and the concrete challenges arising in physics and chemistry, where more manageable approximation schemes are often implemented without paying attention to the subtler points. We try to reconcile rigorous results and practical needs by introducing a family of rougher OIO-type parametrices which involve a well-crafted approximation of the action functional as a Taylor-like expansion.

In this context we present a convergence result in $\mathcal{L}(L^2(\mathbb{R}^d))$ in the spirit of Fujiwara's approach, cf. Theorem 7.4.4. There is of course a price to pay in simplifying: while we retain the same rate of convergence with respect to size of the subdivision $\omega(\Omega)$ of the interval $[s, t]$, the new parametrices are no longer satisfying from the point of view of the semiclassical limit; still, estimates are uniform in $\hbar$ as soon as time is measured in multiples of $\hbar$.

The quest for low-regularity potentials still giving rise to acceptable convergence results for the time-slicing parametrices continues in Chap. 8, that is based on the papers [27, 108, 109, 180]. We consider potentials with at most quadratic growth and regularity level involving suitable families of function spaces that are thoroughly covered in Sect. 3.6.3 below—the so-called Kato-Sobolev spaces [156]. As a rule of thumb, the second derivatives of $V(t, x)$ with respect to x have the regularity of a function in the usual Sobolev space $H^{d+1}(\mathbb{R}^d)$, hence still bounded and continuous, but need not decay at infinity. This is clearly a low-regularity version of Assumption (A) in Sect. 1.3.1 above.

Under this assumption on the potential and the framework of Fujiwara's time-slicing approximation discussed before, we prove a convergence result in the operator norm that is formally identical to the first part of Theorem 1.3.1 for zero-order parametrices, see Theorem 8.4.3. The case of higher-order parametrices is covered in Theorem 8.5.2 under additional regularity assumptions on the potential.

We highlight that the Kato-Sobolev spaces can be regarded as special cases of amalgam spaces of Wiener type, introduced by Feichtinger in [82, 84], illustrating their usefulness in phase-space analysis and certain formal similarities with the modulation spaces. In fact, they turn out to be modulation spaces on the Fourier side, cf. Sect. 3.3 for more details.

We remark that working at this level of regularity makes this chapter the most demanding from the technical point of view. In particular, it is necessary to carefully review the major steps of Fujiwara's construction in this setting. We tried to provide a smooth presentation of the results, often by giving up generality in order to favour the overall fluency of the chapter.

1.3.4 Convergence of Time-Slicing Approximations in the L^p Setting

So far we encountered some convergence results for Feynman path integrals in the L^2 setting. While this is certainly the most natural choice from the point of view of both mathematics and physics, it is quite natural to wonder whether there exists an L^p analogue of Theorem 1.3.1 with $p \neq 2$ under the same assumptions. This is precisely the problem addressed in Chap. 9, motivated by the results that first appeared in [179].

We cannot expect a naive transposition of the claim for several reasons. First of all, notice that the Schrödinger propagator for the free particle is not even bounded on $L^p(\mathbb{R}^d)$ for $p \neq 2$: indeed, the parabolic geometry of its characteristic manifold implies that a peculiar loss of derivative occurs, ultimately due to dispersion, namely [29, 177]

$$\left\| e^{i\hbar\Delta/2} f \right\|_{L^p} \leq C \left\| (1 - h\Delta)^{r/2} f \right\|_{L^p}, \quad r = 2d|1/2 - 1/p|, \quad 1 < p < \infty.$$

It is thus natural to frame the problem in the scale of semiclassical L^p-based Sobolev spaces $H_h^{r,p}(\mathbb{R}^d)$ consisting of all the distributions $f \in \mathcal{S}'(\mathbb{R}^d)$ satisfying

$$\|f\|_{H_h^{r,p}} = \|(1 - h\Delta)^{r/2} f\|_{L^p} < \infty.$$

However, it is not immediately clear how to recast the main ingredients of the time-slicing scheme in this setting; actually, it seems a truly hopeless program: Fujiwara's results highly depend on L^2 estimates for OIOs and composition properties for operator algebras that clearly are no longer satisfied.

The hints collected in the previous discussions, in particular Sect. 1.2, point in one direction: lifting the problem to the phase-space level usually makes things easier. This is precisely what happens here thanks to non-trivial embeddings between the spaces $H_h^{r,p}$ and suitable semiclassical (i.e., $\hbar$-dependent) modulation spaces M_h^p—cf. Theorem 5.1.1. The problem then reduces to a careful analysis of the Gabor matrices of the involved operators, revealing a sparsity phenomenon that is responsible for boundedness and also for a crucial nice behaviour under composition. Coming back to the original domain, we are finally able to prove a convergence result for the time-slicing parametrices under the same regularity assumptions as in Theorem 1.3.1; see Theorem 9.4.1 below for a detailed statement.

In Chap. 9 we also discuss partial extensions in the presence of a magnetic field and also for low-regular potentials in the Sjöstrand class.

We stress that the background framework for Chaps. 8 and 9 is provided by some results obtained by Fujiwara [108, 109, 111]. While the latter are reported and discussed in detail, the proofs are omitted because of the highly technical content that would lead us too far.

Part I
Elements of Gabor Analysis

Chapter 2
Basic Facts of Classical Analysis

The purpose of this chapter is to fix the notation used in the manuscript, as well as to collect the basic facts and main results of real, functional and harmonic analysis that are needed below.

2.1 General Notation

We denote the set of non-negative integer numbers by $\mathbb{N} = \{0, 1, 2, \ldots\}$, while $\mathbb{Z}$, $\mathbb{R}$ and $\mathbb{C}$ are the usual sets of integers, real and complex numbers respectively. In particular, i denotes the imaginary unit and $\bar{z}$ is the complex conjugate of $z \in \mathbb{C}$.

If f is a function from some set A with values in some set B we write $f : A \to B$. We assume to deal with complex-valued functions, i.e. $B = \mathbb{C}$, if not otherwise specified. We write 1_A for the characteristic function of a set A.

We use the symbol $X \lesssim Y$ meaning that the underlying inequality holds up to a universal positive constant factor, namely

$$X \lesssim Y \quad \Longrightarrow \quad \exists C > 0 : X \le CY.$$

If the constant $C = C(a) > 0$ depends on some "allowable" parameter a we write $X \lesssim_a Y$. Moreover, $X \asymp Y$ means that X and Y are *equivalent quantities*, that is both $X \lesssim Y$ and $Y \lesssim X$ hold.

We will be mainly concerned with the d-dimensional real Euclidean space $\mathbb{R}^d$. The standard inner product on $\mathbb{R}^d$ and the induced Euclidean norm are denoted by

$$x \cdot y := \sum_{j=1}^{d} x_j y_j, \quad |x| := \sqrt{x \cdot x} = (x_1^2 + \ldots + x_d^2)^{1/2}.$$

© The Author(s), under exclusive license to Springer Nature Switzerland AG 2022
F. Nicola, S. I. Trapasso, *Wave Packet Analysis of Feynman Path Integrals*,
Lecture Notes in Mathematics 2305, https://doi.org/10.1007/978-3-031-06186-8_2

We write x^2 in place of $|x|^2 = x \cdot x$. Notice that $|x|$ is the absolute value of x in the case where $d = 1$. Recall that $|x|_1 := |x_1| + \ldots + |x_d|$ is an equivalent norm on $\mathbb{R}^d$.

We use several symbols for partial differential operators on $\mathbb{R}^d$:

$$\partial_j = \partial_{x_j} = \frac{\partial}{\partial x_j}, \quad D_j := \frac{1}{2\pi i}\partial_j, \quad j = 1, \ldots, d.$$

We employ the *multi-index notation*: given $\alpha = (\alpha_1, \ldots, \alpha_d) \in \mathbb{N}^d$ and $x \in \mathbb{R}^d$, we write

$$|\alpha| = |\alpha|_1 = \alpha_1 + \ldots + \alpha_d, \quad \alpha! = \alpha_1! \cdots \alpha_d!, \quad x^\alpha = x_1^{\alpha_1} \cdots x_d^{\alpha_d}.$$

$$\partial_x^\alpha = \partial_{x_1}^{\alpha_1} \cdots \partial_{x_d}^{\alpha_d}, \quad D_x^\alpha = D_{x_1}^{\alpha_1} \cdots D_{x_d}^{\alpha_d}.$$

The following relations between multi-indices $\alpha, \beta \in \mathbb{N}^d$ are defined:

$$\alpha \leq \beta \Leftrightarrow \alpha_j \leq \beta_j, \; j = 1, \ldots, d, \qquad \alpha < \beta \Leftrightarrow \alpha \neq \beta \text{ and } \alpha \leq \beta,$$

$$\binom{\alpha}{\beta} \equiv \prod_{j=1}^d \binom{\alpha_j}{\beta_j} = \frac{\alpha!}{\beta!(\alpha - \beta)!} \quad (\text{if } \beta \leq \alpha).$$

We repeatedly make use of the so-called *Japanese brackets* to denote the inhomogeneous magnitude of $x \in \mathbb{R}^d$, namely $\langle x \rangle := (1 + x^2)^{1/2}$. It is useful to remark that the so-called *Peetre inequality* holds:

$$\langle x + y \rangle^s \lesssim_s \langle x \rangle^s \langle y \rangle^{|s|}, \quad x, y \in \mathbb{R}^d, \; s \in \mathbb{R}. \tag{2.1}$$

The set of real matrices of dimension $d \times d$ is denoted by $\mathbb{R}^{d \times d}$. In particular, $I = I_d \in \mathbb{R}^{d \times d}$ is the identity matrix and $O = O_d \in \mathbb{R}^{d \times d}$ is the null matrix. Recall that the canonical symplectic matrix $J \in \mathbb{R}^{2d \times 2d}$ is the nonsingular and skew-symmetric block matrix

$$J = \begin{bmatrix} O & I \\ -I & O \end{bmatrix}.$$

The direct sum $A \oplus B$ of $A, B \in \mathbb{R}^{d \times d}$ is the $2d \times 2d$ matrix defined as

$$A \oplus B = \begin{bmatrix} A & O \\ O & B \end{bmatrix}.$$

Given a list $a_1, \ldots, a_d$ of real numbers, we define the matrix $\text{diag}(a_1, \ldots, a_d) \in \mathbb{R}^{d \times d}$ by

$$[\text{diag}(a_1, \ldots, a_d)]_{ij} := \begin{cases} a_j & (i = j) \\ 0 & (i \neq j), \end{cases}$$

where $[M]_{ij}$ is the element on the i-th row and the j-th column of the matrix $M \in \mathbb{R}^{d \times d}$, $1 \leq i, j \leq d$. Accordingly, we extend the notation to block matrices by setting $\text{diag}(A, B) := A \oplus B$.

We will thoroughly work with invertible matrices, namely elements of the group

$$GL(2d, \mathbb{R}) = \{M \in \mathbb{R}^{2d \times 2d} \mid \det M \neq 0\}.$$

Recall that the *singular values* $s_j(A)$ of a matrix $A \in \mathbb{R}^{d \times d}$ are the positive eigenvalues of $|A| := \sqrt{A^\top A}$, where $A^\top$ denotes the transpose of A; we arrange them in decreasing order: $s_1 \geq s_2 \geq \ldots \geq 0$.

2.2 Function Spaces

We collect below some fundamental results of real analysis and fix the related notation once for all. In what follows we always consider functions $f : \mathbb{R}^d \to \mathbb{C}$, where $\mathbb{R}^d$ is provided with the Lebesgue measure.

2.2.1 Lebesgue Spaces

Let $1 \leq p < \infty$ and $\Omega \subseteq \mathbb{R}^d$ be a measurable set. We consider the class $L^p(\Omega)$ of measurable functions $f : \Omega \to \mathbb{C}$ such that

$$\|f\|_{L^p} := \left(\int_\Omega |f(x)|^p dx \right)^{1/p} < \infty,$$

where as usual two functions are identified if they coincide almost everywhere.

The space $L^\infty(\Omega)$ is defined similarly, where

$$\|f\|_{L^\infty} := \operatorname*{ess\,sup}_{x \in \Omega} |f(x)|.$$

We will be mostly concerned with the case where $\Omega = \mathbb{R}^d$, then we sometimes write L^p for $L^p(\mathbb{R}^d)$.

It will be useful in several occasions to have a more precise and direct control on the integrability of a function; this can be achieved by introducing *weighted Lebesgue spaces*. Given a *weight function*, namely a positive continuous mapping $m: \mathbb{R}^d \to (0, +\infty)$ and $1 \le p \le \infty$ we define the weighted Lebesgue space $L_m^p(\mathbb{R}^d)$ of functions f such that $\|f\|_{L_m^p} := \|f \cdot m\|_{L^p} < \infty$. Weights of particular relevance for our purposes are those of polynomial type, namely

$$v_s(x) := (1 + |x|)^s, \quad s \in \mathbb{R}, \ x \in \mathbb{R}^d.$$

We usually write $L_s^p(\mathbb{R}^d)$ in place of $L_{v_s}^p(\mathbb{R}^d)$. Note that

$$v_s(x) \asymp \langle x \rangle^s \asymp (1 + |x_1| + \ldots + |x_d|)^s, \quad x \in \mathbb{R}^d,$$

hence we will tacitly switch from one form to another one whenever convenient.

For the sake of completeness we list below some basic properties of Lebesgue spaces.

Proposition 2.2.1 *Let m be a weight function on $\mathbb{R}^d$.*

(i) *For any $1 \le p \le \infty$, $L_m^p(\mathbb{R}^d)$ is a Banach space with the norm $\| \cdot \|_{L_m^p}$.*

(ii) *$L^2(\mathbb{R}^d)$ is a Hilbert space with inner product given by*

$$\langle f, g \rangle_{L^2} = \int_{\mathbb{R}^d} f(y)\overline{g(y)}dy.$$

(iii) *(Hölder inequality) For $1 \le p \le \infty$ define the conjugate index p' in such a way that $1/p + 1/p' = 1$—with the understanding that $p' = 1$ if $p = \infty$. If $f \in L_m^p(\mathbb{R}^d)$ and $g \in L_{1/m}^{p'}(\mathbb{R}^d)$ then $fg \in L^1(\mathbb{R}^d)$ and $\|fg\|_{L^1} \le \|f\|_{L_m^p}\|g\|_{L_{1/m}^{p'}}$.*

(iv) *(Duality) If $1 \le p < \infty$ then $L_m^p(\mathbb{R}^d)' = L_{1/m}^{p'}(\mathbb{R}^d)$, the duality being given by*

$$\langle f, g \rangle = \int_{\mathbb{R}^d} f(y)\overline{g(y)}dy,$$

for $f \in L_m^p(\mathbb{R}^d)$, $g \in L_{1/m}^{p'}(\mathbb{R}^d)$.

For $1 \le p, q \le \infty$ and a weight m on $\mathbb{R}^{2d}$, we introduce the weighted *mixed-norm Lebesgue spaces* $L_m^{p,q}(\mathbb{R}^{2d})$, the norm being

$$\|f\|_{L_m^{p,q}} := \left(\int_{\mathbb{R}^d} \left(\int_{\mathbb{R}^d} |f(x,\xi)|^p m(x,\xi)^p dx \right)^{q/p} d\xi \right)^{1/q},$$

with trivial modifications for the cases $p = \infty$ or $q = \infty$.

In the rest of the monograph we will mainly focus on **standard polynomial weights,** namely any of the following two types of functions:

- The global polynomial weight $m = v_s$ on $\mathbb{R}^{2d}$ for some $s \in \mathbb{R}$; we write $L^{p,q}_{v_s}(\mathbb{R}^{2d})$ for clarity in that case. Note that

$$v_s(z) \asymp \langle z \rangle^s \asymp (1 + |x| + |\xi|)^s, \quad z = (x, \xi) \in \mathbb{R}^{2d}.$$

- The tensor product polynomial weight $m(x, \xi) = v_r(x)v_s(\xi)$, $r, s \in \mathbb{R}$; in this case we write $L^{p,q}_{r,s}(\mathbb{R}^{2d})$.

Most of mixed-norm Lebesgue spaces coincide with vector-valued Lebesgue spaces, for instance $L^{p,q}(\mathbb{R}^{2d}) = L^q(\mathbb{R}^d; L^p(\mathbb{R}^d))$ for $1 \leq p, q < \infty$—cf. [230, Lemma 3.8] for a complete discussion in a more general setting. As a result, such spaces enjoy most of the expected properties from the scalar-valued case, cf. [121, Lemma 11.1.2] and [230].

2.2.2 Differentiable Functions and Distributions

Let $C(\mathbb{R}^d) = C^0(\mathbb{R}^d)$ denote the space of continuous maps $\mathbb{R}^d \to \mathbb{C}$. We also write $C_b(\mathbb{R}^d)$ for the space of bounded continuous functions and $C_0(\mathbb{R}^d)$ for the space of continuous functions vanishing at infinity, namely such that $f(x) \to 0$ as $|x| \to +\infty$.

Given $k \in \mathbb{N}$, $k \geq 1$, we recursively define the vector space $C^k(\mathbb{R}^d)$ as the space of differentiable functions $f : \mathbb{R}^d \to \mathbb{C}$ such that $\partial_j f \in C^{k-1}(\mathbb{R}^d)$, $j = 1, \ldots, d$. We similarly define $C^k_b(\mathbb{R}^d)$ as the space of k times continuously differentiable bounded functions on $\mathbb{R}^d$ with bounded derivatives up to k-th order; it is a Banach space with norm

$$\|f\|_{C^k_b} := \max_{|\alpha| \leq k} \sup_{x \in \mathbb{R}^d} |\partial^\alpha f(x)|.$$

We often write $\|f\|_k$ in place of $\|f\|_{C^k_b}$ for the sake of concision.

We denote by $C^\infty(\mathbb{R}^d) := \bigcap_{k \geq 0} C^k(\mathbb{R}^d)$ the space of smooth functions. We also introduce the space $C^\infty_{\geq k}(\mathbb{R}^d)$ of smooth functions with bounded derivatives of any order larger than $k \in \mathbb{N}$, namely

$$C^\infty_{\geq k}(\mathbb{R}^d) := \left\{ f \in C^\infty(\mathbb{R}^d) : |\partial^\alpha f(x)| \leq C_\alpha \quad \forall x \in \mathbb{R}^d, \ \alpha \in \mathbb{N}^d, \ |\alpha| \geq k \right\}.$$

Notice that $C^\infty_b(\mathbb{R}^d) := C^\infty_{\geq 0}(\mathbb{R}^d) = \cap_{k \geq 0} C^k_b(\mathbb{R}^d)$ coincides with the well-known Hörmander class $S^0_{0,0}(\mathbb{R}^d)$ [126, 141]. Recall that the latter is a Fréchet space under the family of seminorms $\{\|\cdot\|_{C_b^k}\}_{k\in\mathbb{N}}$. We emphasize that the class $C_{\geq k}^{\infty}$ is also known as $S_{0,0}^{(k)}$ in microlocal analysis [214].

The set $C_c^k(\mathbb{R}^d)$, $k \in \mathbb{N} \cup \{\infty\}$, is the subspace of functions in $C^k(\mathbb{R}^d)$ with compact support. In particular, the space $C_c^{\infty}(\mathbb{R}^d)$ is often denoted $\mathcal{D}(\mathbb{R}^d)$ after Schwartz [202] and its elements are traditionally known as test functions. Note that $\mathcal{D}(\mathbb{R}^d) = \bigcup_{K\subset\mathbb{R}^d} \mathcal{D}_K(\mathbb{R}^d)$, where $\mathcal{D}_K$ is the space of smooth functions $\mathbb{R}^d \to \mathbb{C}$ supported in the compact subset K of $\mathbb{R}^d$. We consider $\mathcal{D}$ endowed with the inductive limit topology, so that all the embeddings $\mathcal{D}_K(\mathbb{R}^d) \hookrightarrow \mathcal{D}(\mathbb{R}^d)$ are continuous. The resulting topology is locally convex and $\mathcal{D}(\mathbb{R}^d)$ is complete but not metrizable; we refer for instance to [129] for further details.

Recall that the Schwartz class of rapidly decreasing functions $\mathcal{S}(\mathbb{R}^d)$ is the space of smooth functions f on $\mathbb{R}^d$ satisfying

$$p_m(f) := \max_{|\alpha|+|\beta|\leq m} \sup_{x\in\mathbb{R}^d} \left|x^\alpha \partial^\beta f(x)\right| < \infty, \quad \forall m \in \mathbb{N}.$$

It is a Fréchet space with the projective limit topology induced by the family of seminorms $\{p_m\}_{m\in\mathbb{N}}$ and is a dense subset of $L^p(\mathbb{R}^d)$ for any $1 \leq p < \infty$.

The space of *temperate distributions* $\mathcal{S}'(\mathbb{R}^d)$ consists of continuous linear functionals on $\mathcal{S}(\mathbb{R}^d)$, that is $\mathcal{S}'(\mathbb{R}^d) = \mathcal{L}(\mathcal{S}(\mathbb{R}^d), \mathbb{C})$ (cf. Sect. 2.5), and we set

$$\langle f, \phi \rangle = f(\overline{\phi}), \quad f \in \mathcal{S}'(\mathbb{R}^d), \; \phi \in \mathcal{S}(\mathbb{R}^d).$$

The space of distributions $\mathcal{D}'(\mathbb{R}^d)$ is defined similarly, with the Schwartz class replaced by $\mathcal{D}(\mathbb{R}^d)$.

Example 2.2.2

(i) For $1 \leq p \leq \infty$ any p-integrable complex-valued function $f \in L^p(\mathbb{R}^d)$ can be identified with a temperate distribution:

$$\langle f, \phi \rangle = \int_{\mathbb{R}^d} f(y)\overline{\phi(y)}dy, \quad \phi \in \mathcal{S}(\mathbb{R}^d).$$

Notice that this is a further meaning for the brackets $\langle \cdot, \cdot \rangle$.

(ii) Recall that a complex measure μ on $\mathbb{R}^d$ is a $\mathbb{C}$-valued σ-additive map on the Borel σ-algebra $\mathcal{B}_{\mathbb{R}^d}$ of $\mathbb{R}^d$ satisfying $\mu(\emptyset) = 0$. The space $M(\mathbb{R}^d)$ of all complex measures on $\mathbb{R}^d$ is a Banach space with the norm $\|\mu\|_M := |\mu|(\mathbb{R}^d)$, where the total variation $|\mu|: \mathcal{B}_{\mathbb{R}^d} \to [0, +\infty]$ of μ is defined as

$$|\mu|(B) := \sup_{\pi(B)} \sum_{A\in\pi(B)} |\mu(A)|,$$

the supremum being taken over all the partitions $\pi(B)$ of B into a finite number of pairwise disjoint Borel subsets. We have that $M(\mathbb{R}^d) \subset \mathcal{S}'(\mathbb{R}^d)$, since any

$\mu \in M(\mathbb{R}^d)$ can be identified with a temperate distribution:

$$\langle \mu, \phi \rangle = \int_{\mathbb{R}^d} \overline{\phi(y)} d\mu(y), \quad \phi \in S(\mathbb{R}^d).$$

Given a normed linear space of distributions $X \subset S'(\mathbb{R}^d)$, we set

$$X_{\text{comp}} := \{u \in X \ : \ \text{supp} u \text{ is a compact subset of } \mathbb{R}^d\},$$

$$X_{\text{loc}} := \{u \in S'(\mathbb{R}^d) \ : \ \phi u \in X \ \forall \phi \in C_c^\infty(\mathbb{R}^d)\}.$$

2.3 Basic Operations on Functions and Distributions

Consider a function $f : \mathbb{R}^d \to \mathbb{C}$ and let $x, \xi \in \mathbb{R}^d$. The translation and modulation operators T_x and M_ξ are respectively defined as

$$T_x f(y) := f(y - x), \quad M_\xi f(y) := e^{2\pi i y \cdot \xi} f(y), \quad y \in \mathbb{R}^d.$$

Note that such operators do not commute unless $x \cdot \xi \in \mathbb{Z}$, since

$$T_x M_\xi f(y) = e^{-2\pi i x \cdot \xi} M_\xi T_x f(y).$$

Composition of translation and modulation operators are of primary relevance in time-frequency analysis. We fix the ordering once for all and define the *time-frequency shift operator* along $z = (x, \xi) \in \mathbb{R}^{2d}$ as $\pi(z) := M_\xi T_x$.

The reflection operator $\mathcal{I}$ acts on a function $f : \mathbb{R}^d \to \mathbb{C}$ as $\mathcal{I}f(y) := f(-y)$, $y \in \mathbb{R}^d$. We usually write $f^\vee$ in place of $\mathcal{I}f$ for the sake of readability. We also define the involution f^* of f as $f^*(y) := \overline{f(-y)}$, $y \in \mathbb{R}^d$.

Given a matrix $A \in \mathbb{R}^{d \times d}$, the *dilation operator* δ_A acts on a function $f : \mathbb{R}^d \to \mathbb{C}$ as $\delta_A f(y) := f(Ay)$. We also introduce the unitary dilation $D_A f(y) := |\det A|^{1/2} f(Ay)$. If $A = A_\lambda = \lambda I$ for some $\lambda \in \mathbb{R}$ we write δ_λ for δ_{A_λ} and D_λ for D_{A_λ}.

Recall that the *tensor product* of two functions $f, g : \mathbb{R}^d \to \mathbb{C}$ is defined as

$$f \otimes g : \mathbb{R}^{2d} \to \mathbb{C} \ : \ (x, y) \mapsto f \otimes g(x, y) = f(x)g(y).$$

The tensor product $\otimes$ maps $S(\mathbb{R}^d) \times S(\mathbb{R}^d)$ into $S(\mathbb{R}^{2d})$. The tensor product of two temperate distributions $f, g \in S'(\mathbb{R}^d)$ is the distribution $f \otimes g \in S'(\mathbb{R}^{2d})$ acting on any $\Phi \in S(\mathbb{R}^{2d}_{(x,y)})$ as

$$\langle f \otimes g, \Phi \rangle = \langle f, \overline{\langle g, \Phi(x, y) \rangle_y} \rangle_x,$$

meaning that g acts on the section $\Phi(x, \cdot)$ and then f acts on $\langle g, \Phi(x, \cdot) \rangle \in \mathcal{S}(\mathbb{R}_x^d)$. In particular, $f \otimes g$ is the unique temperate distribution such that

$$\langle f \otimes g, \phi_1 \otimes \phi_2 \rangle = \langle f, \phi_1 \rangle \langle g, \phi_2 \rangle, \quad \forall \phi_1, \phi_2 \in \mathcal{S}(\mathbb{R}^d).$$

In conclusion, recall that the complex conjugate of a temperate distribution $f \in \mathcal{S}'(\mathbb{R}^d)$ is denoted by $\overline{f} \in \mathcal{S}'(\mathbb{R}^d)$ and defined by the rule

$$\langle \overline{f}, \phi \rangle = \overline{\langle f, \overline{\phi} \rangle}, \qquad \phi \in \mathcal{S}(\mathbb{R}^d).$$

2.4 The Fourier Transform

The Fourier transform can be initially defined for $f \in L^1(\mathbb{R}^d)$. Precisely, it is the operator $\mathcal{F} \colon L^1(\mathbb{R}^d) \to L^\infty(\mathbb{R}^d)$ defined by

$$\mathcal{F}(f)(\xi) = \hat{f}(\xi) := \int_{\mathbb{R}^d} e^{-2\pi i x \cdot \xi} f(x) dx, \quad \xi \in \mathbb{R}^d.$$

Among the fundamental results on the Fourier transform, we recall:

- the *Riemann-Lebesgue lemma*, that is $\hat{f} \in C_0(\mathbb{R}^d)$ and $\|\hat{f}\|_{L^\infty} \le \|f\|_{L^1}$;
- the *inversion formula*, namely if $f, \hat{f} \in L^1(\mathbb{R}^d)$ then

$$f(x) = \int_{\mathbb{R}^d} e^{2\pi i x \cdot \xi} \hat{f}(\xi) d\xi,$$

  for almost every $x \in \mathbb{R}^d$;
- the *Plancherel theorem*, namely $\mathcal{F} \colon L^1(\mathbb{R}^d) \cap L^2(\mathbb{R}^d) \to L^2(\mathbb{R}^d)$ extends to a unitary operator on $L^2(\mathbb{R}^d)$, hence $\|\hat{f}\|_{L^2} = \|f\|_{L^2}$;
- the *Hausdorff-Young inequality*, namely for any $1 \le p \le 2$ and $f \in L^p(\mathbb{R}^d)$ we have $\|\hat{f}\|_{L^{p'}} \le \|f\|_{L^p}$.

The restriction of $\mathcal{F}$ to $\mathcal{S}(\mathbb{R}^d)$ yields a continuous automorphism that enjoys the usual properties, in particular the inversion formula $\mathcal{F}^{-1} = I\mathcal{F}$. Moreover, the Fourier transform extends by duality to an isomorphism on $\mathcal{S}'(\mathbb{R}^d)$ as follows:

$$\langle \hat{f}, \hat{g} \rangle = \langle f, g \rangle, \qquad f \in \mathcal{S}'(\mathbb{R}^d), \, g \in \mathcal{S}(\mathbb{R}^d).$$

Direct computations show that

$$\mathcal{F} M_\xi = T_\xi \mathcal{F}, \qquad \mathcal{F} T_x = M_{-x} \mathcal{F}, \qquad x, \xi \in \mathbb{R}^d$$

as operators on $\mathcal{S}(\mathbb{R}^d)$ or $L^2(\mathbb{R}^d)$ or $\mathcal{S}'(\mathbb{R}^d)$.

For future convenience we also define the Fourier-Lebesgue spaces $\mathcal{F}L_s^q(\mathbb{R}^d)$, where $1 \leq q \leq \infty$ and $s \in \mathbb{R}$, consisting of distributions $f \in \mathcal{S}'(\mathbb{R}^d)$ such that

$$\|f\|_{\mathcal{F}L_s^q} := \left\|\mathcal{F}^{-1}f\right\|_{L_s^q} < \infty.$$

Finally, the *symplectic Fourier transform* $\mathcal{F}_\sigma$ of a function/temperate distribution F in $\mathbb{R}^{2d}$ is defined as

$$\mathcal{F}_\sigma F(x, \xi) := \mathcal{F}F(J(x, \xi)) = \mathcal{F}F(\xi, -x), \quad (x, \xi) \in \mathbb{R}^{2d}.$$

Note that this is an involution, that is $\mathcal{F}_\sigma(\mathcal{F}_\sigma F) = F$.

2.4.1 Convolution and Fourier Multipliers

The convolution of two functions $f \in L^p(\mathbb{R}^d)$, $1 \leq p \leq \infty$, and $g \in L^1(\mathbb{R}^d)$, is defined as

$$f * g(x) := \int_{\mathbb{R}^d} f(x - y)g(y)dy.$$

Note that $f * g$ is well defined for a.e. $x \in \mathbb{R}^d$.

The convolution of $f \in \mathcal{S}'(\mathbb{R}^d)$ with a Schwartz function $g \in \mathcal{S}(\mathbb{R}^d)$ is defined as the distribution $f * g \in \mathcal{S}'(\mathbb{R}^d)$ such that

$$\langle f * g, \phi \rangle \equiv \langle f, g^* * \phi \rangle, \qquad \forall \phi \in \mathcal{S}(\mathbb{R}^d).$$

In fact, $f * g \in C^\infty(\mathbb{R}^d)$ is a function of polynomial growth together with all its derivatives.

We state some classical results that will be used below.

Proposition 2.4.1

(i) (Young inequality) Let $1 \leq p, q, r \leq \infty$ satisfy $1/p + 1/q = 1 + 1/r$ and $s_1, s_2, s_3 \in \mathbb{R}$ satisfy

$$s_1 + s_3 \geq 0, \quad s_2 + s_3 \geq 0, \quad s_1 + s_2 \geq 0.$$

If $f \in L_{s_1}^p(\mathbb{R}^d)$ and $g \in L_{s_2}^q(\mathbb{R}^d)$, then $f * g \in L_{-s_3}^r(\mathbb{R}^d)$, with

$$\|f * g\|_{L_{-s_3}^r} \lesssim \|f\|_{L_{s_1}^p} \|g\|_{L_{s_2}^q}.$$

(ii) For any $f \in S'(\mathbb{R}^d)$ and $g \in S(\mathbb{R}^d)$:

$$\mathcal{F}(f * g) = \hat{f} \cdot \hat{g}.$$

We highlight that more general Young-type inequalities hold for weighted and mixed-norm spaces, cf. [13, 121].

We moreover define the *Fourier multiplier* with symbol $m \in S'(\mathbb{R}^d)$ to be the linear map

$$m(D)f := \mathcal{F}^{-1}(m \cdot \hat{f}) = \mathcal{F}^{-1}m * f \in S'(\mathbb{R}^d),$$

which is well defined at least for all $f \in S(\mathbb{R}^d)$.

For $1 < p < \infty, r \in \mathbb{R}$, let $H^{r,p} = H^{r,p}(\mathbb{R}^d)$ denote the L^p-based Sobolev space of (fractional) order r, namely the set of distributions $f \in S'(\mathbb{R}^d)$ such that $(1 - \Delta)^{r/2}f \in L^p(\mathbb{R}^d)$, with the norm

$$\|f\|_{H^{r,p}} = \|(1 - \Delta)^{r/2}f\|_{L^p}.$$

Here

$$(1 - \Delta)^{r/2} = m(D)$$

with $m(\xi) = (1 + 4\pi^2|\xi|^2)^{r/2}$. These Banach spaces are known in the literature under a variety of symbols, such as $L^p_r(\mathbb{R}^d)$ or $W^{r,p}(\mathbb{R}^d)$. In the case where $r \in \mathbb{N}$ we have the more familiar, equivalent characterization of $H^{r,p}(\mathbb{R}^d)$ as the space of all $f \in L^p(\mathbb{R}^d)$ which admit weak derivatives $\partial^\alpha f \in L^p(\mathbb{R}^d)$ for any $\alpha \in \mathbb{N}^d$ with $1 \le |\alpha| \le r$, with equivalent norm given by

$$\|f\|_{H^{r,p}} \asymp \sum_{|\alpha| \le r} \|\partial^\alpha f\|_{L^p}.$$

We also set $H^r(\mathbb{R}^d) = H^{r,2}(\mathbb{R}^d), r \in \mathbb{R}$. By the Parseval theorem we therefore have

$$\|f\|_{H^r(\mathbb{R}^d)} = \left(\int_{\mathbb{R}^d} |\hat{f}(\xi)|^2 (1 + 4\pi^2|\xi|^2)^r \, d\xi \right)^{1/2},$$

namely, $H^r(\mathbb{R}^d) = \mathcal{F}L^2_r(\mathbb{R}^d)$. The reader may consult the monographs [167, 211, 225] for further details.

We adopt the usual notation $H^\kappa(B) = H^{\kappa,2}(B), \kappa \in \mathbb{N}$ (κ will always denote a natural number), for the Hilbert spaces corresponding to the L^2-based Sobolev spaces on an open ball $B \subset \mathbb{R}^d$, with the norm

$$\|f\|_{H^\kappa(B)} = \sup_{|\alpha| \le \kappa} \|\partial^\alpha f\|_{L^2(B)}.$$

In the following we will also use the Gagliardo-Nirenberg-Sobolev inequality (see e.g. [184]): if $f \in H^\kappa(B)$ and $f \in L^\infty(B)$ then

$$\|\partial^\alpha f\|_{L^p(B)} \le C \|f\|_{L^\infty(B)}^{1-|\alpha|/\kappa} \|f\|_{H^\kappa(B)}^{|\alpha|/\kappa}, \tag{2.2}$$

for $|\alpha| \le \kappa$, $1/p = |\alpha|/(2\kappa)$, where B is any open ball of radius 1 in $\mathbb{R}^d$, for a constant $C > 0$ independent of B.

Finally, a Sobolev or L^p norm of a vector-valued function is meant as the supremum of the norms of each entry.

2.5 Some More Facts and Notations

Given two Hausdorff topological (complex) vector spaces X and Y, we consider the space $\mathcal{L}(X, Y)$ of all *continuous linear mappings* between X and Y (we write $\mathcal{L}(X)$ if $Y = X$). It can be endowed with different topologies [223], in which cases we write:

1. $\mathcal{L}_b(X, Y)$, if equipped with the topology of *bounded convergence*, that is uniform convergence on bounded subsets of X;
2. $\mathcal{L}_c(X, Y)$, if equipped with the topology of *compact convergence*, that is uniform convergence on compact subsets of X;
3. $\mathcal{L}_s(X, Y)$, if equipped with the topology of *pointwise convergence*, that is uniform convergence on finite subsets of X.

Notice that if $Y = \mathbb{C}$, $\mathcal{L}_b(X, Y) = X_b'$ (the strong dual of X), while $\mathcal{L}_s(X, Y) = X_s'$ (the weak dual of X). It is tacitly assumed that $\mathcal{L}(X, Y) \equiv \mathcal{L}_b(X, Y)$ unless otherwise specified.

We recall that a (complex) Banach algebra A is a Banach space endowed with a product $\star: A \times A \to A$ satisfying

$$\|x \star y\| \le C \|x\| \|y\| \tag{2.3}$$

for every $x, y \in A$ and some $C > 0$. For instance, if X is a (complex) Banach space then $\mathcal{L}(X)$ is a (complex) Banach algebra under composition.

Remark 2.5.1 It is well known that if A is a Banach space with a complex unital algebra structure such that the product $A \times A \to A$ is continuous, then there exists a norm on A, equivalent to the given one, such that *(i)* (2.3) holds with $C = 1$ and *(ii)* the unit has norm equal to 1 (cf. [195, Theorem 10.2]). We tacitly assume this equivalent norm to be the underlying one whenever concerned with a Banach algebra in this book.

Moreover, to unambiguously fix the notation, given a product of elements $a_1, \ldots, a_N$ in a Banach algebra $(A, \star)$, we write

$$\prod_{k=1}^{N} a_k := a_1 \star a_2 \star \ldots \star a_N.$$

This relation is meant to hold even when $(A, \star)$ is a non-commutative algebra. In this case, it is agreed that the notation on the left-hand side exactly designates the ordered product of $a_1, \ldots, a_N$ on the right-hand side.

Finally, let us recall here for later reference a fundamental tool of the theory of PDEs, namely the Gronwall inequality (in integral form)—see [213, Section 1.2] for a proof and further details on variations and applications.

Theorem 2.5.2 *Let* $t_0, t_1 \in \mathbb{R}$, $t_0 < t_1$, *and let* $u \colon [t_0, t_1] \to [0, +\infty)$ *be a continuous function. Assume that there are* $A \geq 0$ *and a continuous function* $B \colon [t_0, t_1] \to [0, +\infty)$ *such that*

$$u(t) \leq A + \int_{t_0}^{t} B(s)u(s)ds, \quad \forall t \in [t_0, t_1].$$

Then the following inequality holds:

$$u(t) \leq A \exp\left(\int_{t_0}^{t} B(s)ds\right), \quad \forall t \in [t_0, t_1]. \tag{2.4}$$

Chapter 3
The Gabor Analysis of Functions

In this chapter we provide an exposition of the main results of time-frequency analysis of functions and distributions. We do not always provide pointers to the literature or proofs, since our presentation is largely inspired by the excellent reference monographs [16, 50, 121]. The reader is invited to consult these references for further details.

3.1 Time-Frequency Representations

3.1.1 The Short-Time Fourier Transform

The first example of a phase-space representation of a function $f \in L^2(\mathbb{R}^d)$ is provided by the *short-time Fourier transform* (STFT), also known as windowed/sliding Fourier transform or Gabor transform. As already mentioned in the introductory Chap. 1, it does ultimately amount to a decomposition of f along the uniform boxes in phase space occupied by the Gabor atoms $\pi(z)g$, $z \in \mathbb{R}^{2d}$, for some fixed window function $g \in L^2(\mathbb{R}^d) \setminus \{0\}$. Precisely, it is defined as

$$V_g f(x, \xi) := \langle f, \pi(x, \xi)g \rangle = \int_{\mathbb{R}^d} e^{-2\pi i y \cdot \xi} f(y) \overline{g(y - x)} \, dy, \quad (x, \xi) \in \mathbb{R}^{2d}.$$

The following equivalent representations are readily proved:

$$V_g f(x, \xi) = \mathcal{F}(f \cdot \overline{T_x g})(\xi) = e^{-2\pi i x \cdot \xi}(f * M_\xi g^*)(x). \qquad (3.1)$$

Note that while we assumed $g \neq 0$ in the definition of the STFT, that is natural in view of the related motivational discussions, this condition is not essential from the mathematical point of view. The case $g = 0$ is clearly of little significance,

© The Author(s), under exclusive license to Springer Nature Switzerland AG 2022
F. Nicola, S. I. Trapasso, *Wave Packet Analysis of Feynman Path Integrals*,
Lecture Notes in Mathematics 2305, https://doi.org/10.1007/978-3-031-06186-8_3

but still admissible, and the same holds for most of the properties given below. For this reason we usually do not explicitly require the condition $g \neq 0$ to hold, unless strictly necessary.

We stress that if $f \in L^p(\mathbb{R}^d)$, $1 \leq p \leq \infty$, and $g \in L^{p'}(\mathbb{R}^d)$, then $V_g f$ is defined pointwise on $\mathbb{R}^{2d}$. In general, $V_g f$ is defined pointwise by duality if $f \in S'(\mathbb{R}^d)$ and $g \in S(\mathbb{R}^d)$, and also if $f \in X'$ and $g \in X$ for any Banach space X such that $S(\mathbb{R}^d) \hookrightarrow X$ with dense inclusion (so that $X' \hookrightarrow S'(\mathbb{R}^d)$) and such that X is invariant under time-frequency shifts.

In any case where the STFT is defined in some sense, notice that the mapping $(f, g) \mapsto V_g f$ is sesquilinear—namely linear in f and conjugate-linear in g. In particular, for a fixed window g, the map $V_g : f \mapsto V_g f$ is linear.

As far as the regularity of the STFT is concerned, we have the following result.

Proposition 3.1.1

(i) *If $f \in L^p(\mathbb{R}^d)$, $1 < p < \infty$, and $g \in L^{p'}(\mathbb{R}^d)$ then $V_g f \in C_0(\mathbb{R}^{2d})$ and*

$$\|V_g f\|_\infty \leq \|f\|_{L^p} \|g\|_{L^{p'}}.$$

(ii) *If $f \in S'(\mathbb{R}^d)$ and $g \in S(\mathbb{R}^d)$ then $V_g f \in C^\infty(\mathbb{R}^{2d})$ and there exist $N \in \mathbb{N}$ and $C > 0$ such that*

$$|V_g f(z)| \leq C(1 + |z|)^N, \quad z \in \mathbb{R}^{2d}.$$

Moreover $V_g : S'(\mathbb{R}^d) \to S'(\mathbb{R}^{2d})$ is continuous.

(iii) *Fix $g \in S(\mathbb{R}^d) \setminus \{0\}$. If $f \in S'(\mathbb{R}^d)$ then $f \in S(\mathbb{R}^d)$ if and only if $V_g f \in S(\mathbb{R}^{2d})$, which is also equivalent to say that for all $N \geq 0$ there exists $C_N > 0$ such that*

$$|V_g f(z)| \leq C_N(1 + |z|)^{-N}, \quad z \in \mathbb{R}^{2d}.$$

Moreover $V_g : S(\mathbb{R}^d) \to S(\mathbb{R}^{2d})$ is continuous.

The following fundamental properties of the STFT will be used below.

Proposition 3.1.2

(i) *(The fundamental STFT identity)*

$$V_g f(x, \xi) = e^{-2\pi i x \cdot \xi} V_{\hat{g}} \hat{f}(\xi, -x),$$

for all $f, g \in L^2(\mathbb{R}^d)$ and $(x, \xi) \in \mathbb{R}^{2d}$.

(ii) *(Switching identity) If $f, g \in L^2(\mathbb{R}^d)$, then $V_g f(x, \xi) = e^{-2\pi i x \cdot \xi} \overline{V_f g(-x, -\xi)}$, $(x, \xi) \in \mathbb{R}^{2d}$.*

(iii) *(Covariance property) Fix $(u, v) \in \mathbb{R}^{2d}$. Then*

$$V_g(\pi(u, v)f)(x, \xi) = e^{-2\pi i u \cdot (\xi - v)} V_g f(x - u, \xi - v),$$

for all $f, g \in L^2(\mathbb{R}^d)$ and $(x, \xi) \in \mathbb{R}^{2d}$.

(iv) *(Orthogonality relations) If $f_1, f_2 \in L^2(\mathbb{R}^d)$ and $g_1, g_2 \in L^2(\mathbb{R}^d)$, then $V_{g_i} f_i \in L^2(\mathbb{R}^{2d})$, $i = 1, 2$, and*

$$\langle V_{g_1} f_1, V_{g_2} f_2 \rangle_{L^2(\mathbb{R}^{2d})} = \langle f_1, f_2 \rangle_{L^2(\mathbb{R}^d)} \overline{\langle g_1, g_2 \rangle_{L^2(\mathbb{R}^d)}}.$$

(v) *(Adjoint STFT) The adjoint STFT with window $g \in \mathcal{S}(\mathbb{R}^d)$ is the operator V_g^* which maps a measurable function $F: \mathbb{R}^{2d} \to \mathbb{C}$ of polynomial growth into the temperate distribution*

$$V_g^* F = \int_{\mathbb{R}^{2d}} F(z) \pi(z) g \, dz, \qquad (3.2)$$

namely $\langle V_g^* F, \phi \rangle = \langle F, V_g \phi \rangle$, $\phi \in \mathcal{S}(\mathbb{R}^d)$. If $F \in \mathcal{S}(\mathbb{R}^{2d})$ then $V_g^* F \in \mathcal{S}(\mathbb{R}^d)$.

(vi) *(Inversion formula) Let $f \in \mathcal{S}'(\mathbb{R}^d)$ and $g, \gamma \in \mathcal{S}(\mathbb{R}^d)$ be such that $\langle g, \gamma \rangle \neq 0$. Then,*

$$f = \frac{1}{\langle \gamma, g \rangle} V_\gamma^* V_g f = \frac{1}{\langle \gamma, g \rangle} \int_{\mathbb{R}^{2d}} V_g f(z) \pi(z) \gamma \, dz. \qquad (3.3)$$

In particular, the STFT is injective on $\mathcal{S}'(\mathbb{R}^d)$.

(vii) *(Change-of-window formula) Let $g, h, \gamma \in \mathcal{S}(\mathbb{R}^d)$ be such that $\langle h, \gamma \rangle \neq 0$. Then, for any $f \in \mathcal{S}'(\mathbb{R}^d)$,*

$$|V_g f(z)| \leq \frac{1}{|\langle h, \gamma \rangle|} (|V_h f| * |V_g \gamma|)(z), \quad z \in \mathbb{R}^{2d}.$$

(viii) *(Tensor product) Let $f_1, f_2 \in \mathcal{S}'(\mathbb{R}^d)$ and $g \in \mathcal{S}(\mathbb{R}^{2d})$ be such that $g = g_1 \otimes g_2$ for some $g_1, g_2 \in \mathcal{S}(\mathbb{R}^d)$. Then*

$$V_{g_1 \otimes g_2}(f_1 \otimes f_2)(z, \zeta) = V_{g_1} f_1(z_1, \zeta_1) V_{g_2} f_2(z_2, \zeta_2),$$

for any $z = (z_1, z_2) \in \mathbb{R}^{2d}$, $\zeta = (\zeta_1, \zeta_2) \in \mathbb{R}^{2d}$.

Proof We give a sketch of the proof for the benefit of the reader, who is invited to fill in the details as an instructive exercise—full proofs can be found in [121].

(i) follows from the Plancherel theorem:

$$V_g f(x, \xi) = \langle f, M_\xi T_x g \rangle$$

$$= \langle \hat{f}, \mathcal{F}(M_\xi T_x g) \rangle$$

$$= \langle \hat{f}, T_\xi M_{-x} \hat{g} \rangle$$

$$= e^{-2\pi i x \cdot \xi} \langle \hat{f}, M_{-x} T_\xi \hat{g} \rangle,$$

hence the claim.

(ii) and (iii) are direct consequences of the very definition of the STFT and linear changes of variables.

(iv) follows from the Plancherel theorem and Fubini's theorem:

$$\langle V_{g_1} f_1, V_{g_2} f_2 \rangle_{L^2(\mathbb{R}^{2d})} = \int_{\mathbb{R}^d} \langle \mathcal{F}(f_1 \overline{T_x g_1}), \mathcal{F}(f_2 \overline{T_x g_2}) \rangle_{L^2(\mathbb{R}^d)} \, dx$$

$$= \int_{\mathbb{R}^{2d}} f_1(y) \overline{g_1(y - x)} \overline{f_2(y)} g_2(y - x) \, dy \, dx$$

$$= \langle f_1, f_2 \rangle_{L^2(\mathbb{R}^d)} \overline{\langle g_1, g_2 \rangle_{L^2(\mathbb{R}^d)}}$$

(the case $f_1 = f_2$, $g_2 = g_2$ implies the first part of the statement).

(v) means that $V_g^* F$ is well defined as a temperate distribution via

$$\langle V_g^* F, \phi \rangle = \int_{\mathbb{R}^{2d}} F(z) \langle \pi(z) g, \phi \rangle \, dz$$

$$= \int_{\mathbb{R}^{2d}} F(z) \overline{V_g \phi(z)} \, dz$$

$$= \langle F, V_g \phi \rangle,$$

which is clear from Proposition 3.1.1 (iii). The last part of the statement follows by a direct inspection of the above definition of $V_g^* F$ by a formal integral that, if $|F(z)| \leq C_N (1 + |z|)^{-N}$ for every $N \in \mathbb{N}$, actually converges absolutely to a Schwartz function, say $\tilde{f} \in \mathcal{S}(\mathbb{R}^d)$, which satisfies $\langle \tilde{f}, \phi \rangle = \langle F, V_g \phi \rangle$ for every $\phi \in \mathcal{S}(\mathbb{R}^d)$.

Concerning (vi), for $f \in \mathcal{S}'(\mathbb{R}^d)$ and $g, \gamma, \phi \in \mathcal{S}(\mathbb{R}^d)$ we have, by (v),

$$\langle V_\gamma^* V_g f, \phi \rangle = \langle V_g f, V_\gamma \phi \rangle = \langle \gamma, g \rangle \langle f, \phi \rangle,$$

where the last equality follows from (iv) above if $f \in \mathcal{S}(\mathbb{R}^d)$ and for $f \in \mathcal{S}'(\mathbb{R}^d)$ by density, using Proposition 3.1.1 (ii).

For the estimate in (vii) we use the inversion formula above and we obtain

$$f = \frac{1}{\langle \gamma, h \rangle} V_\gamma^* V_h f.$$

Therefore, taking the Gabor transform and using the properties in (ii), (iii) and (v), for any $z = (x, \xi) \in \mathbb{R}^{2d}$ we have

$$V_g f(x, \xi) = \frac{1}{\langle \gamma, h \rangle} V_g V_\gamma^* V_h f(x, \xi)$$

$$= \frac{1}{\langle \gamma, h \rangle} \langle V_\gamma^* V_h f, \pi(x, \xi) g \rangle$$

$$= \frac{1}{\langle \gamma, h \rangle} \langle V_h f, V_\gamma(\pi(x, \xi) g) \rangle$$

$$= \frac{1}{\langle \gamma, h \rangle} \int_{\mathbb{R}^{2d}} e^{-2\pi i u \cdot (\xi - v)} V_h f(u, v) V_g \gamma(x - u, \xi - v) \, du \, dv.$$

The claimed pointwise estimate thus follows after taking absolute values.

Finally, (viii) is a direct consequence of the definition of tensor product of distributions. $\square$

It is easy to see that the definition of $V_g^* F$, $g \in \mathcal{S}(\mathbb{R}^d)$, extends naturally to $F \in \mathcal{S}'(\mathbb{R}^{2d})$, and defines continuous mappings $V_g^* : \mathcal{S}'(\mathbb{R}^{2d}) \to \mathcal{S}'(\mathbb{R}^d)$ and $V_g^* : \mathcal{S}(\mathbb{R}^{2d}) \to \mathcal{S}(\mathbb{R}^d)$.

3.1.2 Quadratic Representations

It is useful in several settings to produce time-frequency representations that depend quadratically on the signal. For instance, similar functions are interpreted as phase-space energy densities in signal analysis or as quasi-probability distributions in quantum physics.

A standard way to obtain a quadratic representation from a sesquilinear one, say $L(f, g)$, is to consider $Q(f) = L(f, f)$—possibly up to complex factors. For any $\alpha_1, \alpha_2 \in \mathbb{C}$ and suitable functions f_1, f_2 we have

$$Q(\alpha_1 f_1 + \alpha_2 f_2) = |\alpha_1|^2 Q(f_1) + \alpha_1 \overline{\alpha_2} L(f_1, f_2) + \overline{\alpha_1} \alpha_2 L(f_2, f_1) + |\alpha_2|^2 Q(f_2).$$

The non-linear behaviour due to the occurrence of interference cross-terms is distinctive of quadratic representations. It is a major drawback both for applications and theoretical purposes, but it is somehow compensated by other nice features as showed below.

The most common quadratic representations (also called quadratic functions, distributions or transforms) in signal analysis are the *spectrogram* and the *radar ambiguity distribution*, which are derived from the short-time Fourier transform as described above. In particular, the spectrogram of the signal $f \in L^2(\mathbb{R}^d)$ with respect to a window $g \in L^2(\mathbb{R}^d)$ such that $\|g\|_{L^2} = 1$ is defined by $S_g f(z) := |V_g f(z)|^2$, $z \in \mathbb{R}^{2d}$ (in this case $L(f_1, f_2) = V_g f_1 \overline{V_g f_2}$).

Observe that $V_g f(z)$ is not equal to $\overline{V_f g(z)}$, and $V_f f$ is not real-valued, in general. This is clearly due to the imbalance of the STFT with respect to the role played by f and g. In light of these remarks, it could be tempting to fix these issues by "distributing" the effect of the space translation as suggested by the identity

$$V_g f(x, \xi) = e^{-\pi i x \cdot \xi} \langle T_{-x/2} f, M_\xi T_{x/2} g \rangle.$$

The latter expression motivates the introduction of the cross-ambiguity function of functions $f, g \in L^2(\mathbb{R}^d)$ as

$$\text{Amb}(f, g)(x, \xi) := \langle T_{-x/2} f, M_\xi T_{x/2} g \rangle = \int_{\mathbb{R}^d} e^{-2\pi i y \cdot \xi} f(y + x/2) \overline{g(y - x/2)} dy$$

with $(x, \xi) \in \mathbb{R}^{2d}$, and the corresponding ambiguity function of $f \in L^2(\mathbb{R}^d)$ as

$$\text{Amb} f(x, \xi) := \text{Amb}(f, f)(x, \xi) = \int_{\mathbb{R}^d} e^{-2\pi i y \cdot \xi} f(y + x/2) \overline{f(y - x/2)} dy.$$

Hence we have

$$\text{Amb} f(x, \xi) = e^{\pi i x \cdot \xi} V_f f(x, \xi), \quad \text{Amb}(f, g)(x, \xi) = e^{\pi i x \cdot \xi} V_g f(x, \xi).$$

It is then clear that most of the properties of the short-time Fourier transform are inherited by the spectrogram and the ambiguity function, therefore we omit an explicit description here.

The symplectic Fourier transform of $\text{Amb}(f, g)$ is the so-called *cross-Wigner distribution* of $f, g \in L^2(\mathbb{R}^d)$, which is easily seen to be given by

$$W(f, g)(x, \xi) := \int_{\mathbb{R}^d} e^{-2\pi i y \cdot \xi} f(x + y/2) \overline{g(x - y/2)} dy, \quad (x, \xi) \in \mathbb{R}^{2d},$$

the corresponding *Wigner distribution/transform* (WD) of $f \in L^2(\mathbb{R}^d)$ being

$$W f(x, \xi) := W(f, f)(x, \xi) = \int_{\mathbb{R}^d} e^{-2\pi i y \cdot \xi} f(x + y/2) \overline{f(x - y/2)} dy, \quad (x, \xi) \in \mathbb{R}^{2d}.$$

It is not difficult to show that the Wigner transform is related to the STFT as follows:

$$W(f, g)(x, \xi) = 2^d e^{4\pi i x \cdot \xi} V_{g^\vee} f(2x, 2\xi), \quad (x, \xi) \in \mathbb{R}^{2d}.$$

As already observed,

$$W(f, g) = \mathcal{F}_\sigma \text{Amb}(f, g), \quad \text{Amb}(f, g) = \mathcal{F}_\sigma W(f, g).$$

As a result of these connections, most of the properties of the STFT in Propositions 3.1.1 and 3.1.2 extend to the Wigner distribution. Nevertheless, there are some features which are quite distinctive of this representation; we list below those which are used in the rest of this monograph.

Proposition 3.1.3 *Let* $g, g_1, g_2, f, f_1, f_2 \in L^2(\mathbb{R}^d)$.

(i) *We have*

$$W(f, g) = \mathcal{F}_2 \mathfrak{T}_s (f \otimes \overline{g}),$$

where $\mathcal{F}_2$ *is the partial Fourier transform with respect to the second variable on* $\mathbb{R}^{2d}$ *and* $\mathfrak{T}_s$ *acts on* $F: \mathbb{R}^{2d} \to \mathbb{C}$ *as*

$$\mathfrak{T}_s F(x, y) = F\left(\frac{x+y}{2}, x - y\right).$$

These operators are automorphisms of $\mathcal{S}(\mathbb{R}^{2d})$, *extend to automorphisms of* $\mathcal{S}'(\mathbb{R}^{2d})$ *and allow one to define* $W(f, g)$ *for* $f, g \in \mathcal{S}'(\mathbb{R}^d)$.

(ii) *(Schwartz regularity) If* $f, g \in \mathcal{S}(\mathbb{R}^d)$ *then* $W(f, g) \in \mathcal{S}(\mathbb{R}^{2d})$ *and the correspondence* $\mathcal{S}(\mathbb{R}^d) \times \mathcal{S}(\mathbb{R}^d) \ni (f, g) \mapsto W(f, g) \in \mathcal{S}(\mathbb{R}^{2d})$ *is continuous.*

(iii) *(The fundamental WD identity)* $W(f, g)(x, \xi) = W(\hat{f}, \hat{g})(\xi, -x)$.

(iv) *(Real-valuedness)* $W(f, g)(z) = \overline{W(g, f)(z)}$. *In particular,* $Wf: \mathbb{R}^{2d} \to \mathbb{R}$.

(v) *(Covariance property) For every* $u, v, z \in \mathbb{R}^{2d}$,

$$W(\pi(u)f, \pi(v)g)(z) = e^{\pi i(v_1+u_1)\cdot(v_2-u_2)} M_{J(u-v)} T_{\frac{u+v}{2}} W(f, g)(z). \quad (3.4)$$

In particular,

$$Wf(\pi(u)f)(z) = Wf(z - u).$$

(vi) *(Moyal's formula)* $W(f_i, g_i) \in L^2(\mathbb{R}^d)$, $i = 1, 2$, *and*

$$\langle W(f_1, g_1), W(f_2, g_2) \rangle = \langle f_1, f_2 \rangle \overline{\langle g_1, g_2 \rangle}. \quad (3.5)$$

In particular, $\|Wf\|_{L^2} = \|f\|_{L^2}^2$.

(vii) *(Marginal densities) If* $f, g \in \mathcal{S}(\mathbb{R}^d)$ *then*

$$\int_{\mathbb{R}^d} W(f, g)(x, \xi)d\xi = f(x)\overline{g(x)}, \quad \int_{\mathbb{R}^d} W(f, g)(x, \xi)dx = \hat{f}(\xi)\overline{\hat{g}(\xi)}.$$

(viii) *(Projective identification) If $Wf = Wg$ then there exists $c \in \mathbb{C}$, $|c| = 1$, such that $f = cg$.*

(ix) *(The "magic formula" [119]) Let $\phi \in S(\mathbb{R}^d)$ and set $\Phi = W\phi \in S(\mathbb{R}^{2d})$. For all $z = (z_1, z_2), \zeta = (\zeta_1, \zeta_2) \in \mathbb{R}^{2d}$,*

$$V_\Phi W(g, f)(z, \zeta) = e^{-2\pi i z_2 \cdot \zeta_2} \overline{V_\phi f(z_1 + \zeta_2/2, z_2 - \zeta_1/2)}$$
$$\times V_\phi g(z_1 - \zeta_2/2, z_2 + \zeta_1/2).$$

3.2   Modulation Spaces

Modulation spaces were introduced by H. Feichtinger in the early 1980s [83, 86]. According to the inventor [87], the original motivation in defining modulation spaces is rooted in the theory of harmonic analysis over locally compact Abelian groups; in particular, the need of a whole family of Banach spaces closed under duality and complex interpolation lead to the definition of modulation spaces in terms of uniform decompositions on the spectral side of their members. Therefore, in the first instance they can be thought of as Besov spaces with isometric boxes in the frequency domain instead of dyadic annuli. A much more insightful definition is given in terms of the global decay/summability of the phase-space concentration of a function or a distribution; this is in fact the so-called *coorbit representation* of modulation spaces, which falls in the perspective of the general coorbit theory developed by Feichtinger and Gröchenig around 1990 [91–93].

Definition 3.2.1 Let m be a standard polynomial weight on $\mathbb{R}^{2d}$—that is a weight of the form $m(z) = v_s(z), s \in \mathbb{R}$, or $m(z) = v_r(x)v_s(\xi)$, for $r, s \in \mathbb{R}$ and $z = (x, \xi)$. Let $1 \le p, q \le \infty$ and fix $g \in S(\mathbb{R}^d) \setminus \{0\}$. The modulation space $M_m^{p,q}(\mathbb{R}^d)$ is the set of all temperate distributions $f \in S'(\mathbb{R}^d)$ such that

$$\|f\|_{M_m^{p,q}} := \|V_g f\|_{L_m^{p,q}} = \left(\int_{\mathbb{R}^d} \left(\int_{\mathbb{R}^d} |V_g f(x, \xi)|^p m(x, \xi)^p dx \right)^{q/p} d\xi \right)^{1/q} < \infty,$$

(3.6)

with suitable modifications in the cases $p = \infty$ or $q = \infty$. If $p = q$ we write $M_m^p(\mathbb{R}^d)$; we omit the subscript for the weight if $m \equiv 1$ and write $M^p(\mathbb{R}^d)$. Moreover, if $m = v_r \otimes v_s$ we write $M_{r,s}^{p,q}(\mathbb{R}^d)$. If there is no risk of confusion we usually write $M_m^{p,q}$ for $M_m^{p,q}(\mathbb{R}^d)$.

Remark 3.2.2

(i) The name "modulation space" comes from noticing that $|V_g f(x, \xi)| = |f * M_\xi g^*(x)|$. The test function g is thus deformed by means of modulation, while in context of Besov spaces one is ultimately concerned with the Lebesgue regularity of $f * \delta_\xi g$, where $\delta_\xi g$ is a suitable dilation. Such characterizations were used by Peetre and Triebel, and lead to develop the theory of modulation spaces by paralleling that of the Besov spaces, cf. [186, 224].

(ii) The definition of modulation spaces can be given in more general settings. A definition of $M^{p,q}$ in the case $0 < p, q \le \infty$ (quasi-Banach setting) was provided by Galperin and Samarah in [114], while mixed modulation spaces were also considered; see e.g. [49]. Vector-valued modulation spaces were first studied by Wahlberg in [230].

(iii) We restrict to polynomial weights both for simplicity and also in order to frame the theory of modulation spaces in the context of temperate distributions. Nevertheless, more general weights of temperate growth can be taken into account in the same setting—cf. Sect. 3.6.1 below and [121, 218], while weight functions with growth faster than polynomial at infinity may lead to enlarge the universe space to ultra-distributions [123, 216].

We collect in the following result the main properties of modulation spaces that will be used throughout the monograph.

Proposition 3.2.3 *Let m, m_1, m_2 be standard polynomial weight functions on $\mathbb{R}^{2d}$.*

(i) *(Banach space property) For any $1 \le p, q \le \infty$, the modulation space $M_m^{p,q}(\mathbb{R}^d)$ is a Banach space with the norm (3.6) which is invariant under time-frequency shifts. The definition does not depend on the window function g in (3.6), in the sense that $\left\| V_\phi f \right\|_{L_m^{p,q}}$ is an equivalent norm for $M_m^{p,q}$ for any choice of $\phi \in S(\mathbb{R}^d) \setminus \{0\}$.*

(ii) *(Density of S) The Schwartz space $S(\mathbb{R}^d)$ is a subset of $M_m^{p,q}(\mathbb{R}^d)$ for any $1 \le p, q \le \infty$, in particular a dense subset if $p, q \ne \infty$. Moreover, the following characterization holds for any $1 \le p, q \le \infty$:*

$$S(\mathbb{R}^d) = \bigcap_{s \ge 0} M_{v_s}^{p,q}(\mathbb{R}^d), \quad S'(\mathbb{R}^d) = \bigcup_{s \ge 0} M_{v_s}^{p,q}(\mathbb{R}^d).$$

(iii) *(Reconstruction formula) If $g, \gamma \in S(\mathbb{R}^d)$, then $V_g : M_m^{p,q}(\mathbb{R}^d) \to L_m^{p,q}(\mathbb{R}^{2d})$ and $V_\gamma^* : L_m^{p,q}(\mathbb{R}^{2d}) \to M_m^{p,q}(\mathbb{R}^d)$ are bounded operators. Moreover, if $\langle g, \gamma \rangle \ne 0$ then the inversion formula (3.3) holds for any $f \in M_m^{p,q}(\mathbb{R}^d)$; in short, $\mathrm{Id}_{M_m^{p,q}(\mathbb{R}^d)} = \langle \gamma, g \rangle^{-1} V_\gamma^* V_g$.*

(iv) *(Duality) If $1 \le p, q < \infty$ then $(M_m^{p,q}(\mathbb{R}^d))' \simeq M_{1/m}^{p',q'}(\mathbb{R}^d)$ and the duality pairing is given by*

$$\langle f, h \rangle = \int_{\mathbb{R}^{2d}} V_g f(z) \overline{V_g h(z)} dz,$$

for $f \in M_m^{p,q}(\mathbb{R}^d)$, $h \in M_{1/m}^{p',q'}(\mathbb{R}^d)$ and $g \in \mathcal{S}(\mathbb{R}^d) \setminus \{0\}$. As a consequence, for any $1 < p, q \leq \infty$,

$$\|f\|_{M_m^{p,q}} = \sup_{h \in M_{1/m}^{p',q'}} |\langle f, h \rangle|.$$

(v) (Inclusions) If $p_1 \leq p_2$, $q_1 \leq q_2$ and $m_1 \gtrsim m_2$ then $M_{m_1}^{p_1,q_1}(\mathbb{R}^d) \subset M_{m_2}^{p_2,q_2}(\mathbb{R}^d)$. In particular, for any $1 \leq p, q \leq \infty$,

$$M^1(\mathbb{R}^d) \subset M^{p,q}(\mathbb{R}^d) \subset M^\infty(\mathbb{R}^d).$$

(vi) (Local properties) For any $1 \leq p, q \leq \infty$,

$$(M^{p,q}(\mathbb{R}^d))_{\mathrm{loc}} = (\mathcal{F}L^q(\mathbb{R}^d))_{\mathrm{loc}}, \quad (M^{p,q}(\mathbb{R}^d))_{\mathrm{comp}} = (\mathcal{F}L^q(\mathbb{R}^d))_{\mathrm{comp}}.$$

(vii) (Complex interpolation) Let $0 < \theta < 1$, $1 \leq p_1, p_2, q_1, q_2 \leq \infty$. Then $[M_{m_1}^{p_1,q_1}(\mathbb{R}^d), M_{m_2}^{p_2,q_2}(\mathbb{R}^d)]_{[\theta]} = M_m^{p,q}(\mathbb{R}^d)$, where

$$\frac{1}{p} = \frac{1-\theta}{p_1} + \frac{\theta}{p_2}, \quad \frac{1}{q} = \frac{1-\theta}{q_1} + \frac{\theta}{q_2}, \quad m = m_1^{1-\theta} m_2^\theta.$$

The boundedness of some operations on modulation spaces is established in the following result.

Proposition 3.2.4

(i) (Convolution) Let $1 \leq p, q, p_1, q_1, p_2, q_2 \leq \infty$. Then

$$\|f * g\|_{M^{p,q}} \lesssim \|f\|_{M^{p_1,q_1}} \|g\|_{M^{p_2,q_2}}$$

if and only if

$$\frac{1}{p_1} + \frac{1}{p_2} \geq 1 + \frac{1}{p}, \quad \frac{1}{q_1} + \frac{1}{q_2} \geq \frac{1}{q}.$$

(ii) (Multiplication) Let $1 \leq p, q, p_1, q_1, p_2, q_2 \leq \infty$. Then

$$\|f \cdot g\|_{M^{p,q}} \lesssim \|f\|_{M^{p_1,q_1}} \|g\|_{M^{p_2,q_2}}$$

if and only if

$$\frac{1}{p_1} + \frac{1}{p_2} \geq \frac{1}{p}, \quad \frac{1}{q_1} + \frac{1}{q_2} \geq 1 + \frac{1}{q}.$$

(iii) *(Dilation) Let $A \in \mathrm{GL}(d, \mathbb{R})$ and $1 \leq p, q \leq \infty$. For any $f \in M^{p,q}(\mathbb{R}^d)$,*

$$\|\delta_A f\|_{M^{p,q}} \lesssim C_{p,q}(A) \|f\|_{M^{p,q}},$$

where $C_{p,q}(A) = |\det A|^{-(1/p+1/q')}(\det(I + A^\top A))^{1/2}$.

(iv) *(Tensor product) Let m_i be standard polynomial weights and $f_i \in M_{m_i}^{p,q}(\mathbb{R}^d)$, $i = 1, 2$. Then*

$$\|f_1 \otimes f_2\|_{M_m^{p,q}} \lesssim \|f_1\|_{M_{m_1}^{p,q}} \|f_2\|_{M_{m_2}^{p,q}}, \tag{3.7}$$

where $m(z, \zeta) := m_1(z_1, \zeta_1) m_2(z_2, \zeta_2)$ for $z = (z_1, z_2) \in \mathbb{R}^{2d}$ and $\zeta = (\zeta_1, \zeta_2) \in \mathbb{R}^{2d}$.

We wish to emphasize that many common function spaces are embedded in modulation spaces.

Proposition 3.2.5

(i) *If $s \in \mathbb{R}$ then $M_{s,0}^2(\mathbb{R}^d)$ coincides with the weighted Lebesgue space $L_s^2(\mathbb{R}^d)$, while $M_{0,s}^2(\mathbb{R}^d)$ coincides with the Sobolev space $H^s(\mathbb{R}^d)$.*

(ii) *The following continuous embeddings with Lebesgue spaces hold:*

$$M^{p,q_1}(\mathbb{R}^d) \subset L^p(\mathbb{R}^d) \subset M^{p,q_2}(\mathbb{R}^d), \quad q_1 \leq \min\{p, p'\}, \quad q_2 \geq \max\{p, p'\}.$$

Similarly,

$$M^{q_1,p}(\mathbb{R}^d) \subset \mathcal{F} L^p(\mathbb{R}^d) \subset M^{q_2,q}(\mathbb{R}^d), \quad q_1 \leq \min\{2, p'\}, \quad q_2 \geq \max\{2, p'\}.$$

Moreover, the following embedding results (see [160, Theorems 1.3-1.4] and also [161]) will be crucial in the following.

Theorem 3.2.6 *Let $1 < p < \infty$ and $r = 2d|1/2 - 1/p|$. Then we have*

$$M^p(\mathbb{R}^d) \hookrightarrow L^p(\mathbb{R}^d), \quad H^{r,p}(\mathbb{R}^d) \hookrightarrow M^p(\mathbb{R}^d), \quad 1 < p \leq 2 \tag{3.8}$$

as well as

$$L^p(\mathbb{R}^d) \hookrightarrow M^p(\mathbb{R}^d), \quad M^p(\mathbb{R}^d) \hookrightarrow H^{-r,p}(\mathbb{R}^d), \quad 2 \leq p < \infty.$$

3.3   Wiener Amalgam Spaces

Let us consider the action of the Fourier transform on $f \in M_m^{p,q}(\mathbb{R}^d)$, where $1 \leq p,q \leq \infty$ and m is a standard polynomial weight. Recall from Proposition 3.1.2 that, for $g \in \mathcal{S}(\mathbb{R}^d)$,

$$|V_g f(z)| = |V_{\hat{g}} \hat{f}(Jz)|,$$

hence

$$\|f\|_{M_m^{p,q}} = \left(\int_{\mathbb{R}^d} \left(\int_{\mathbb{R}^d} |V_{\hat{g}} \hat{f}(x,\xi)|^p m(-\xi,x)^p d\xi \right)^{q/p} dx \right)^{1/q}.$$

Therefore, the space $\mathcal{F} M_m^{p,q}(\mathbb{R}^d)$ comes with a Banach norm given by

$$\|h\|_{\mathcal{F} M_m^{p,q}} = \left(\int_{\mathbb{R}^d} \left(\int_{\mathbb{R}^d} |V_g h(x,\xi)|^p m(\xi,-x)^p d\xi \right)^{q/p} dx \right)^{1/q}.$$

We see that this norm is similar to that of the corresponding modulation space (3.6), but with reversed order of integration over time and frequency and a swap of variables in the weight function.

We define in general the space $W_m^{p,q}(\mathbb{R}^d)$ as $W_m^{p,q}(\mathbb{R}^d) := \mathcal{F} M_{m \circ J}^{p,q}(\mathbb{R}^d)$. The related notation is ultimately identical to that used for modulation spaces, for instance we write $W^p(\mathbb{R}^d)$ if $p = q$ and $m \equiv 1$, and $W_{r,s}^{p,q}(\mathbb{R}^d)$ if $m = v_r \otimes v_s$, $r, s \in \mathbb{R}$.

The special connection between these families of spaces provides other interesting relations, such as the following set of generalized Hausdorff-Young inequalities—which are a direct consequence of Minkowski's integral inequality:

$$M_{r,s}^{p,q}(\mathbb{R}^d) \hookrightarrow W_{s,r}^{q,p}(\mathbb{R}^d), \quad 1 \leq q \leq p \leq \infty, \, r, s \in \mathbb{R}. \tag{3.9}$$

$$W_{r,s}^{p,q}(\mathbb{R}^d) \hookrightarrow M_{s,r}^{p,q}(\mathbb{R}^d), \quad 1 \leq p \leq q \leq \infty, \, r, s \in \mathbb{R}.$$

As a result, note that $M^p(\mathbb{R}^d) = W^p(\mathbb{R}^d)$ for any $1 \leq p \leq \infty$, in particular $M^2(\mathbb{R}^d) = W^2(\mathbb{R}^d) = L^2(\mathbb{R}^d)$.

It is worth highlighting that for $f \in M_{r,s}^{p,q}(\mathbb{R}^d)$ we have

$$\|\hat{f}\|_{W_{r,s}^{p,q}} = \|\|\hat{f} \cdot \overline{T_x \hat{g}}(\xi)\|_{\mathcal{F} L_r^p(\mathbb{R}_\xi^d)}\|_{L_s^q(\mathbb{R}_x^d)} = \|f\|_{M_{r,s}^{p,q}}.$$

Therefore, the norm of the space $W_{r,s}^{p,q}$ is defined by imposing a global L^q-regularity condition on a "sliding" local $\mathcal{F}L^p$-regularity measure. This structure in fact characterizes the norm of *Wiener amalgam spaces*, where global and local features of functions are "amalgamated". The seminal papers on the topic date back to the work of N. Wiener [235–237] and H. Feichtinger [82, 84]; for a systematic review of the subject we address to the references [31, 85, 107, 132, 139]. We emphasize that Wiener amalgam spaces were introduced before modulation spaces and in fact Feichtinger admitted that the latter were originally designed as amalgam spaces on the Fourier side [87].

Even if the definition of amalgam spaces can be given in greater generality, we consider only those which are relevant to the purposes of this monograph.

Definition 3.3.1 Let $(B, \|\cdot\|_B)$ any of the spaces $L_u^p(\mathbb{R}^d)$, $\mathcal{F}L_u^p(\mathbb{R}^d)$ or $L_m^{p,q}(\mathbb{R}^{2d})$ for some $1 \leq p, q \leq \infty$ and standard polynomial weights u on $\mathbb{R}^d$ and m on $\mathbb{R}^{2d}$. Let $(C, \|\cdot\|_C)$ be any of the spaces $L_u^p(\mathbb{R}^d)$ or $L_m^{p,q}(\mathbb{R}^{2d})$ for some $1 \leq p, q \leq \infty$. The Wiener amalgam space $W(B, C)$ with local component B and global component C is defined as

$$W(B, C)(\mathbb{R}^d) = \{f \in B_{\text{loc}} : F_g(f) \in C\},$$

where $g \in C_c^\infty(\mathbb{R}^d) \setminus \{0\}$ is a fixed window function and F_g is the associated control function: $F_g(f)(x) := \|f \cdot T_x g\|_B$. The natural norm on $W(B, C)$ is $\|f\|_{W(B,C)} := \|F_g\|_C$.

Note in particular that $W_{r,s}^{p,q}(\mathbb{R}^d) = W(\mathcal{F}L_r^p, L_s^q)(\mathbb{R}^d)$. Another class of Wiener amalgam spaces that will play an important role in Chap. 8 is that of *Kato-Sobolev spaces*, also known as uniformly local Sobolev spaces, introduced by T. Kato in [156]. Precisely, for $r \in \mathbb{R}$ the space $H_{\text{ul}}^r(\mathbb{R}^d)$ consists of all the distributions $f \in \mathcal{S}'(\mathbb{R}^d)$ such that

$$\|f\|_{H_{\text{ul}}^r} := \sup_{y \in \mathbb{R}^d} \|g(\cdot - y)f\|_{H^r} < \infty,$$

where $g \in C_c^\infty(\mathbb{R}^d) \setminus \{0\}$ and $H^r(\mathbb{R}^d) = \mathcal{F}L_r^2(\mathbb{R}^d)$ is the L^2-based Sobolev space of order r (cf. Sect. 2.4). We thus recognize that

$$H_{\text{ul}}^r(\mathbb{R}^d) = W(H^r, L^\infty)(\mathbb{R}^d) = W(\mathcal{F}L_r^2, L^\infty)(\mathbb{R}^d) = W_{r,0}^{2,\infty}(\mathbb{R}^d) = \mathcal{F}M_{r,0}^{2,\infty}(\mathbb{R}^d).$$

We refer to Sect. 3.6.3 for a detailed account on this class of functions with local Sobolev regularity and boundedness at infinity—see also [7] for a comprehensive discussion. While the properties proved in that section could be recast using results from the theory of amalgam spaces, we prefer to provide direct arguments for the benefit of the reader.

We now state the main basic properties of Wiener amalgam spaces.

Proposition 3.3.2 *Let B, B_i be local component spaces and C, C_i be global component spaces as in Definition 3.3.1, $i = 1, 2, 3$.*

  (i) *(Banach space property) $(W(B, C)(\mathbb{R}^d), \| \cdot \|_{W(B,C)})$ is a Banach space continuously embedded into $S'(\mathbb{R}^d)$; moreover, different window functions in $C_c^\infty(\mathbb{R}^d)$ provide equivalent norms.*
 (ii) *(Embeddings) If $B_1 \hookrightarrow B_2$ and $C_1 \hookrightarrow C_2$, then $W(B_1, C_1) \hookrightarrow W(B_2, C_2)$.*
(iii) *(Complex interpolation) For $0 < \theta < 1$ we have*

$$[W(B_1, C_1), W(B_2, C_2)]_{[\theta]} = W([B_1, B_2]_{[\theta]}, [C_1, C_2]_{[\theta]})$$

  if C_1 or C_2 has absolutely continuous norm.[1]
 (iv) *(Duality) If B', C' are the topological dual spaces of B, C respectively, and the space of test functions C_c^∞ is dense in both B and C, then $W(B, C)' = W(B', C')$.*
  (v) *(Admissible windows) The class of admissible windows for the norm in the definition of $W^{p,q}(\mathbb{R}^d)$, $1 \le p, q \le \infty$, can be extended from $C_c^\infty(\mathbb{R}^d)$ to $M^1(\mathbb{R}^d) = W^1(\mathbb{R}^d)$.*

Note that this general framework allows one to use the tools of the theory of decomposition spaces to further study the properties of such spaces. In order to exploit this connection we resort to an equivalent discrete norm for the amalgam spaces $W^{p,q}(\mathbb{R}^d)$; the general case is covered in [84]. Recall that a *bounded uniform partition of function* (BUPU) $(\{\psi_i\}_{i \in I}, (x_i)_{i \in I}, U)$ consists of a family of non-negative functions $\{\psi_i\}_{i \in I}$ in $\mathcal{F}L^1(\mathbb{R}^d)$ such that the following conditions are satisfied:

1. $\sum_{i \in I} \psi_i(x) = 1$, for any $x \in \mathbb{R}^d$;
2. $\sup_{i \in I} \|\psi_i\|_{\mathcal{F}L^1} < \infty$;
3. there exist a discrete family $(x_i)_{i \in I}$ in $\mathbb{R}^d$ and a relatively compact set $U \subset \mathbb{R}^d$ such that $\mathrm{supp}(\psi_i) \subset x_i + U$ for any $i \in I$, and
4. $\sup_{i \in I} \#\{j : x_i + U \cap x_j + U \ne \emptyset\} < \infty$.

A general result in the theory of amalgam spaces is the following norm equivalence in the spirit of decomposition spaces:

$$\|f\|_{W^{p,q}} \asymp \left(\sum_{i \in I} \|f\, \psi_i\|_{\mathcal{F}L^p}^q \right)^{1/q}. \tag{3.10}$$

---

[1] Recall that a Banach space $(Y, \| \|_Y)$ of functions $\mathbb{R}^n \to \mathbb{C}$ has an absolutely continuous norm if dominated convergence holds for all $f \in Y$; more precisely, if $\{f_n\} \subset Y$ is such that $|f_n(t)| \le |g(t)|$ for a.e. $t \in \mathbb{R}^n$ and some $g \in Y$, and $f_n(t) \to f(t)$ for a.e. $t \in \mathbb{R}^n$ then $\|f_n - f\|_Y \to 0$. See [14] for further details.

A similar characterization holds for modulation spaces [86, 231], providing a norm comparable to that of Besov spaces:

$$\|f\|_{M^{p,q}} \asymp \left(\sum_{i \in I} \|\Box_i f\|_{L^p}^q \right)^{1/q}$$

where we introduced the frequency-uniform decomposition operators

$$\Box_i := \mathcal{F}^{-1} \psi_i \mathcal{F}, \qquad i \in I.$$

In particular, a Young type result for amalgam spaces can be obtained, see [84, Theorem 3]—and [50, 131, 221] for more general convolution and multiplication properties of weighted Wiener amalgam spaces and modulation spaces.

Theorem 3.3.3 *For any* $1 \le p_1, p_2, p_3, q_1, q_2, q_3 \le \infty$ *such that*

$$\mathcal{F}L^{p_1}(\mathbb{R}^d) * \mathcal{F}L^{p_2}(\mathbb{R}^d) \hookrightarrow \mathcal{F}L^{p_3}(\mathbb{R}^d),$$

$$L^{q_1}(\mathbb{R}^d) * L^{q_2}(\mathbb{R}^d) \hookrightarrow L^{q_3}(\mathbb{R}^d),$$

namely $1/p_1 + 1/p_2 = 1/p_3$, $1/q_1 + 1/q_2 = 1/q_3 + 1$, *the following inclusion holds:*

$$W^{p_1,q_1}(\mathbb{R}^d) * W^{p_2,q_2}(\mathbb{R}^d) \hookrightarrow W^{p_3,q_3}(\mathbb{R}^d).$$

Proof For the benefit of the reader we sketch here a short proof in the spirit of [132, Theorem 11.8.3]. We consider as a BUPU for $W^{p,q}(\mathbb{R}^d)$ the family $\{\psi_k\}_{k \in \mathbb{Z}^d} \subset C_c^\infty(\mathbb{R}^d) \subset \mathcal{F}L^1(\mathbb{R}^d)$ defined by

$$\psi_k(t) = \frac{\phi(t-k)}{\sum_{k \in \mathbb{Z}^d} \phi(t-k)}, \qquad t \in \mathbb{R}^d,$$

for a fixed $\phi \in C_c^\infty(\mathbb{R}^d)$, $\phi \ge 0$, such that $\phi(t) = 1$ for $t \in [0,1]^d$ and $\phi(t) = 0$ for $t \in \mathbb{R}^d \setminus [-1, 2]^d$. After introducing the control functions

$$\Psi_{f,p}(k) := \|f\,\psi_k\|_{\mathcal{F}L^p}, \qquad k \in \mathbb{Z}^d,$$

the equivalent norm (3.10) becomes

$$\|f\|_{W^{p,q}} \asymp \left(\sum_{k \in \mathbb{Z}^d} \|f\,\psi_k\|_{\mathcal{F}L^p}^q \right)^{1/q} = \|\Psi_{f,p}\|_{\ell^q(\mathbb{Z}^d)}.$$

54   3   The Gabor Analysis of Functions

For $f \in W^{p_1,q_1}$ and $g \in W^{p_2,q_2}$ set $f_m = f\psi_m$, $g_n = g\psi_n$ for $m, n \in \mathbb{Z}^d$. Hence we have

$$\text{supp}(f_m * g_n) \subset \text{supp}(f_m) + \text{supp}(g_n) = m + n + 2\,\text{supp}\psi.$$

It is then clear that the cardinality of the set

$$J_k := \{(m, n) \in \mathbb{Z}^{2d} : \text{supp}((f_m * g_n)\psi_k) \neq \emptyset\}$$

is finite for any $k \in \mathbb{Z}^d$ and is uniformly bounded with respect to m, n, k. In fact,

$$J_k = \{(m, n) \in \mathbb{Z}^{2d} : m = k - n + \alpha, \ |\alpha| \leq N(d)\},$$

for a fixed constant $N(d) \in \mathbb{N}$ depending only on the dimension d. Therefore, an easy computation yields

$$\Psi_{f*g,p_3}(k) = \sum_{|\alpha| \leq N(d)} \Psi_{f,p_1} * \Psi_{g,p_2}(k + \alpha),$$

and thus

$$\|f * g\|_{W^{p_3,q_3}} \lesssim \|f\|_{W^{p_1,q_1}} \|g\|_{W^{p_2,q_2}},$$

that is the claim. $\square$

Remark 3.3.4 In view of the relation with modulation spaces, under the same assumptions of the previous theorem we also have

$$M^{p_1,q_1}(\mathbb{R}^d) \cdot M^{p_2,q_2}(\mathbb{R}^d) \hookrightarrow M^{p_3,q_3}(\mathbb{R}^d).$$

Remark 3.3.5 Theorem 3.3.3 is a special case of a general stability result under convolution and pointwise product for Wiener amalgam spaces, that we state here for completeness—see [84, 90] for proofs. Let B_1, B_2, B_3 and C_1, C_2, C_3 be local and global component spaces respectively, as in Definition 3.3.1.

(i) If $B_1 \cdot B_2 \hookrightarrow B_3$ and $C_1 \cdot C_2 \hookrightarrow C_3$, then

$$W(B_1, C_1) \cdot W(B_2, C_2) \hookrightarrow W(B_3, C_3).$$

(ii) If $B_1 * B_2 \hookrightarrow B_3$ and $C_1 * C_2 \hookrightarrow C_3$, then

$$W(B_1, C_1) * W(B_2, C_2) \hookrightarrow W(B_3, C_3).$$

3.4 A Banach-Gelfand Triple of Modulation Spaces

The modulation space $M^1(\mathbb{R}^d)$ has a rather special role in the theory of modulation spaces. It is in fact the first kind of modulation space introduced by Feichtinger as a special, new Segal algebra [83]. It is known since then as the Feichtinger algebra; we address the reader to the recent paper [148] for a comprehensive survey on the topic.

Note that $M^1(\mathbb{R}^d) \subset M^{p,q}(\mathbb{R}^d) \subset M^\infty(\mathbb{R}^d)$ for any $1 \le p, q \le \infty$, and in particular $M^1(\mathbb{R}^d) \subset L^2(\mathbb{R}^d) \subset M^\infty(\mathbb{R}^d)$. There has been an increasing interest for the triple (M^1, M^2, M^∞) as a replacement of the standard Gelfand triple $(\mathcal{S}, L^2, \mathcal{S}')$ of real harmonic analysis [98]. Recall that a *Gelfand triple* (also known as rigged Hilbert space in quantum mechanics) consists of a separable Hilbert space H and a topological vector space Q such that the inclusion $i : Q \to H$ is an injective continuous operator with dense image [6]; note that the adjoint map $i^* : H \simeq H' \to Q'$ is injective too and the inner product on H extends to the duality pairing $Q - Q'$ in a natural way.

In particular, Feichtinger's algebra is a good substitute of the Schwartz class of test functions for the purposes of Gabor analysis—except for applications to PDEs or to situations where a control on the regularity is required [89, 95]. We list below the main properties of M^1 as a function space—it may be helpful to recall that $M^1 = W^1$.

Proposition 3.4.1

(i) $f \in M^1(\mathbb{R}^d)$ if and only if $f \in L^2(\mathbb{R}^d)$ and $V_g f \in L^1(\mathbb{R}^{2d})$ for all $g \in M^1(\mathbb{R}^d)$. In particular $f \in M^1(\mathbb{R}^d) \iff f \in L^2(\mathbb{R}^d)$ and $\mathrm{Amb} f \in L^1(\mathbb{R}^{2d}) \iff f \in L^2(\mathbb{R}^d)$ and $Wf \in L^1(\mathbb{R}^{2d})$.

(ii) If $f \in M^1(\mathbb{R}^d)$, then $f, \hat{f} \in L^1(\mathbb{R}^d)$ and f is continuous.

(iii) $M^1(\mathbb{R}^d)$ is a time-frequency homogeneous Banach space: for any $z \in \mathbb{R}^{2d}$, $f \in M^1(\mathbb{R}^d)$, one has $\pi(z)f \in M^1(\mathbb{R}^d)$ and $\|\pi(z)f\|_{M^1} = \|f\|_{M^1}$. In particular, it is the smallest time-frequency homogeneous Banach space containing the Gaussian function.

(iv) The Schwartz class $\mathcal{S}(\mathbb{R}^d)$ is a dense subset of $M^1(\mathbb{R}^d)$ and $L^2(\mathbb{R}^d)$ is the completion of $M^1(\mathbb{R}^d)$ with respect to the $L^2(\mathbb{R}^d)$ norm.

(v) $M^1(\mathbb{R}^d)$ is invariant under the Fourier transform, i.e. for any $f \in M^1(\mathbb{R}^d)$ one has $\hat{f} \in M^1(\mathbb{R}^d)$ and $\|\hat{f}\|_{M^1} = \|f\|_{M^1}$.

(vi) $M^1(\mathbb{R}^d)$ is a Banach algebra under both convolution and pointwise multiplication.

(vii) The class of admissible windows for computing the norm of the modulation space $M^{p,q}(\mathbb{R}^d)$ can be extended to $M^1(\mathbb{R}^d)$, meaning that $\|V_\phi f\|_{L^{p,q}}$ is an equivalent norm for $M^{p,q}$, $1 \le p, q \le \infty$, for any choice of $\phi \in M^1(\mathbb{R}^d)\backslash\{0\}$.

The Feichtinger algebra is also well behaved under tensor products and we list a few results in this respect.

Proposition 3.4.2

(i) *The tensor product* $\otimes \colon M^1(\mathbb{R}^d) \times M^1(\mathbb{R}^d) \to M^1(\mathbb{R}^{2d})$ *is a bilinear bounded operator.*

(ii) M^1 *enjoys the projective tensor factorization property:*

$$M^1(\mathbb{R}^{2d}) \simeq M^1(\mathbb{R}^d) \widehat{\otimes} M^1(\mathbb{R}^d),$$

namely the space $M^1(\mathbb{R}^{2d})$ *consists of all functions of the form*

$$f = \sum_{n \in \mathbb{N}} g_n \otimes h_n,$$

where $\{g_n\}, \{h_n\}$ *are (sequences of) functions in* $M^1(\mathbb{R}^d)$ *such that*

$$\sum_{n \in \mathbb{N}} \|g_n\|_{M^1} \|h_n\|_{M^1} < \infty.$$

(iii) *The tensor product is well defined on* $M^\infty(\mathbb{R}^d)$: *for any* $f, g \in M^\infty(\mathbb{R}^d)$, $f \otimes g$ *is the unique element of* $M^\infty(\mathbb{R}^{2d})$ *such that*

$$\langle f \otimes g, \phi_1 \otimes \phi_2 \rangle \equiv \langle f, \phi_1 \rangle \langle g, \phi_2 \rangle, \quad \forall \phi_1, \phi_2 \in M^1(\mathbb{R}^d).$$

The parallel with the standard triple $(\mathcal{S}, L^2, \mathcal{S}')$ is further reinforced by a fundamental kernel theorem; just as the (temperate) distributions are related to the Schwartz kernel theorem, the *Feichtinger kernel theorem* [94] characterizes operators $M^1 \to M^\infty$ as follows—see also [49].

Theorem 3.4.3

(i) *Every distribution* $k \in M^\infty(\mathbb{R}^{2d})$ *defines a bounded linear operator* $T_k \colon M^1(\mathbb{R}^d) \to M^\infty(\mathbb{R}^d)$ *according to the rule*

$$\langle T_k f, g \rangle = \langle k, g \otimes \overline{f} \rangle, \quad \forall f, g \in M^1(\mathbb{R}^d),$$

with $\|T_k\|_{M^1 \to M^\infty} \lesssim \|k\|_{M^\infty}$.

(ii) *For any bounded operator* $T \colon M^1(\mathbb{R}^d) \to M^\infty(\mathbb{R}^d)$ *there exists a unique kernel* $k_T \in M^\infty(\mathbb{R}^{2d})$ *such that*

$$\langle Tf, g \rangle = \langle k_T, g \otimes \overline{f} \rangle, \quad \forall f, g \in M^1(\mathbb{R}^d).$$

3.5 The Sjöstrand Class and Related Spaces

Another special member of the family of modulation spaces is $M^{\infty,1}(\mathbb{R}^d)$. This is also known as the Sjöstrand class after the seminal paper [209], where it was introduced as an exotic symbol class still yielding bounded pseudodifferential operators on $L^2(\mathbb{R}^d)$. It was later recognized that operators with symbols in this space enjoy several other properties.

Pseudodifferential operators with Sjöstrand symbols have also been used as potential perturbations for the Schrödinger equation, since they are particularly well suited to the Gabor analysis of the corresponding propagator—cf. Chap. 4 below.

We highlight here some important properties that will be used below.

Proposition 3.5.1

(i) $M^{\infty,1}(\mathbb{R}^d) \subset (\mathcal{F}L^1(\mathbb{R}^d))_{\text{loc}} \cap L^\infty(\mathbb{R}^d) \subset C_b(\mathbb{R}^d)$. *More precisely,* $M^{\infty,1}(\mathbb{R}^d) \subset W^{1,\infty}(\mathbb{R}^d)$.

(ii) $(M^{\infty,1}(\mathbb{R}^d))_{\text{loc}} = (\mathcal{F}L^1(\mathbb{R}^d))_{\text{loc}}$.

(iii) *If* $k \in \mathbb{N}$ *and* $k > d$ *then* $C_b^k(\mathbb{R}^d) \subset M^{\infty,1}(\mathbb{R}^d)$. *Moreover,*

$$C_b^\infty(\mathbb{R}^d) = \bigcap_{s \geq 0} M_{0,s}^\infty(\mathbb{R}^d) = \bigcap_{s \geq 0} M_{0,s}^{\infty,1}(\mathbb{R}^d).$$

(iv) $\mathcal{F}M(\mathbb{R}^d) \subset M^{\infty,1}(\mathbb{R}^d)$, *where* $M(\mathbb{R}^d)$ *is the space of complex measures on* $\mathbb{R}^d$.

Proof

(i) It is a direct consequence of the definition. The refined inclusion $M^{\infty,1}(\mathbb{R}^d) \subset W^{1,\infty}(\mathbb{R}^d)$ follows from (3.9).

(ii) See Proposition 3.2.3—for the proof, see [16, Proposition 2.9].

(iii) The proof can be found in [121, Theorem 14.5.3] and [126, Lemma 6.1].

(iv) We prove equivalently that $M(\mathbb{R}^d) \subset W^{\infty,1}(\mathbb{R}^d)$. Arguing in terms of Wiener amalgam spaces, since $M(\mathbb{R}^d) = W(M, L^1)(\mathbb{R}^d)$ by [88, Theorem 1] we can resort to [82, Remark 1.2] to equivalently conclude that $M(\mathbb{R}^d) = W(M, L^1)(\mathbb{R}^d) \subset W(\mathcal{F}L^\infty, L^1)(\mathbb{R}^d) = \mathcal{F}M^{\infty,1}(\mathbb{R}^d)$.

A direct proof can be provided as well. Recall that $M(\mathbb{R}^d) \subset \mathcal{S}'(\mathbb{R}^d)$, hence for any non-zero window $g \in \mathcal{S}(\mathbb{R}^d)$ we can explicitly compute the STFT of $\mu \in M(\mathbb{R}^d)$:

$$V_g\mu(x, \xi) = \langle \mu, M_\xi T_x g \rangle = \int_{\mathbb{R}^d} e^{-2\pi i y \cdot \xi} \overline{g(y - x)} d\mu(y).$$

Therefore,

$$\|\mu\|_{W^{\infty,1}} = \int_{\mathbb{R}^d} \sup_{\xi \in \mathbb{R}^d} |V_g\mu(x,\xi)| dx$$

$$\leq \int_{\mathbb{R}^d} \sup_{\xi \in \mathbb{R}^d} \int_{\mathbb{R}^d} |e^{-2\pi i y \cdot \xi} \overline{g(y-x)}| d|\mu|(y)\, dx$$

$$= \int_{\mathbb{R}^d} \int_{\mathbb{R}^d} |g(y-x)| d|\mu|(y)\, dx$$

$$= \int_{\mathbb{R}^d} \int_{\mathbb{R}^d} |g(y-x)| dx\, d|\mu|(y)$$

$$= \|g\|_{L^1} |\mu|(\mathbb{R}^d) < \infty,$$

as claimed.

□

Remark 3.5.2 For the sake of concreteness we emphasize that any function $f = g * h$ obtained by smoothing a distribution $h \in M^{\infty}(\mathbb{R}^d)$ with a convolution filter $g \in M^1(\mathbb{R}^d)$ belongs to $M^{\infty,1}(\mathbb{R}^d)$. This is readily proved arguing on the spectral side in light of Remark 3.3.5, since

$$\hat{f} = \hat{g} \cdot \hat{h} \in W^1(\mathbb{R}^d) \cdot W^{\infty}(\mathbb{R}^d) \subset W^{\infty,1}(\mathbb{R}^d).$$

This gives us the opportunity to build many interesting examples in a quite simple fashion.

Consider for instance functions which are obtained via periodization of a given $g \in M^1(\mathbb{R}^d)$ along a regular lattice $\Lambda = \alpha\mathbb{Z}^d$ with $\alpha > 0$, that is $f(x) = g * \text{III}_\Lambda(x) := \sum_{k \in \mathbb{Z}^d} g(x - \alpha k)$, where we introduced the Dirac comb $\text{III}_\Lambda := \sum_{k \in \mathbb{Z}^d} \delta_{\alpha k} \in M^{\infty}(\mathbb{R}^d)$. We have indeed that $f \in M^{\infty,1}(\mathbb{R}^d)$ and the same holds for $f = g * h$ with $h = \sum_{k \in \mathbb{Z}} v_k \delta_{\alpha k}$ for any $(v_k)_{k \in \mathbb{Z}} \in \ell^{\infty}(\mathbb{Z})$, since $h \in W(\mathcal{M}, \ell^{\infty})(\mathbb{R}) \subset M^{\infty}(\mathbb{R})$ as well.

As a concrete example, in dimension $d = 1$, any continuous piecewise linear function f with values $(v_k)_{k \in \mathbb{Z}} \in \ell^{\infty}(\mathbb{Z})$ at α-separated integer nodes $\{\alpha k\}_{k \in \mathbb{Z}} \subset \mathbb{R}$, for some $\alpha > 0$, can be written as $f = g * h$ as above, with $g(x) = \max\{1 - |x|/\alpha, 0\}$ (triangular function, which belongs to $\mathcal{F}L^1(\mathbb{R})$ and is compactly supported, hence in $M^1(\mathbb{R})$) and $h = \sum_{k \in \mathbb{Z}} v_k \delta_{\alpha k} \in M^{\infty}(\mathbb{R})$.

It turns out that the Sjöstrand class can be equipped with a Banach algebra structure with respect to the pointwise multiplication, as a consequence of the following characterization [192, Theorem 3.5 and Corollary 2.10]—in fact, it also enjoys a non-commutative algebra structure under Weyl product of symbols, see Sect. 4.2 below.

Proposition 3.5.3 *Let* $1 \leq p, q \leq \infty$ *and* $s \in \mathbb{R}$. *The following facts are equivalent.*

(i) $M_{0,s}^{p,q}(\mathbb{R}^d)$ *is a Banach algebra for pointwise multiplication.*
(ii) $M_{0,s}^{p,q}(\mathbb{R}^d) \hookrightarrow L^\infty(\mathbb{R}^d)$.
(iii) *Either* $s = 0$ *and* $q = 1$ *or* $s > d/q'$.

Remark 3.5.4 Let us stress that the aforementioned results concern conditions under which the embedding $M_{0,s}^{p,q} \cdot M_{0,s}^{p,q} \hookrightarrow M_{0,s}^{p,q}$ is continuous, hence there exists a constant $C > 0$ such that

$$\|fg\|_{M_{0,s}^{p,q}} \leq C \|f\|_{M_{0,s}^{p,q}} \|g\|_{M_{0,s}^{p,q}}, \qquad \forall f, g \in M_{0,s}^{p,q}(\mathbb{R}^d).$$

As a result, the Banach algebra property holds here only up to a constant. Nevertheless, we already agreed in Remark 2.5.1 that we tacitly assume the norm to be normalized in such a way that $C = 1$.

Functions in $M^{\infty,1}$ enjoy nice low-pass decompositions, in the sense of the following results.

Proposition 3.5.5 *For any* $\epsilon > 0$ *and* $f \in M^{\infty,1}(\mathbb{R}^d)$, *there exist* $f_1 \in C_b^\infty(\mathbb{R}^d)$ *and* $f_2 \in M^{\infty,1}(\mathbb{R}^d)$ *such that*

$$f = f_1 + f_2, \qquad \|f_2\|_{M^{\infty,1}} \leq \epsilon.$$

Proof Fix $g \in \mathcal{S}(\mathbb{R}^d)$ with $\|g\|_{L^2} = 1$, and set

$$f_1(y) = V_g\left(V_g f \cdot 1_{A_R}\right)(y) = \int_{A_R} V_g f(x, \xi) e^{2\pi i y \cdot \xi} g(y - x) dx d\xi, \qquad (3.11)$$

in the sense of distributions, where $A_R = \{(x, \xi) \in \mathbb{R}^{2d} : |\xi| \leq R\}$, and $R > 0$ will be chosen later, depending on ϵ.

The integral in (3.11) in fact converges for every y and defines a bounded function. Indeed, setting $S(\xi) = \sup_{x \in \mathbb{R}^d} |V_g f(x, \xi)|$, we have $S \in L^1(\mathbb{R}^d)$ by the assumption $f \in M^{\infty,1}(\mathbb{R}^d)$, and for any $y \in \mathbb{R}^d$,

$$|f_1(y)| \leq \int_{A_R} |V_g f(x, \xi)| |g(y - x)| dx d\xi$$

$$\leq \|g\|_{L^1} \|S\|_{L^1}.$$

It can be proved similarly that all the derivatives $\partial^\alpha f_1$ are bounded, using that $\xi^\alpha S(\xi)$ is integrable on $|\xi| \leq R$. Differentiation under the integral sign is allowed, because for y in a neighbourhood of any fixed $y_0 \in \mathbb{R}^d$ and every N,

$$|V_g f(x, \xi) \partial_y^\alpha [e^{2\pi i y \cdot \xi} g(y - x)]| \leq C_N (1 + |\xi|)^{|\alpha|} S(\xi)(1 + |y_0 - x|)^{-N},$$

which is integrable in A_R. Hence $f_1 \in C_b^\infty(\mathbb{R}^d)$.

Let us now consider $f_2 = f - f_1 = V_g(V_g f \cdot 1_{A_R^c})$, where the second equality entails the inversion formula for the STFT (3.3). The continuity of $V_g^* : L^{\infty,1}(\mathbb{R}^{2d}) \to M^{\infty,1}(\mathbb{R}^d)$ yields

$$
\begin{aligned}
\|f_2\|_{M^{\infty,1}} &= \|V_g(V_g f \cdot 1_{A_R^c})\|_{M^{\infty,1}} \\
&\lesssim \|V_g f \cdot 1_{A_R^c}\|_{L^{\infty,1}} \\
&= \int_{|\xi|>R} S(\xi) d\xi \leq \epsilon
\end{aligned}
$$

provided that $R = R_\epsilon$ is large enough.   □

3.6   Complements

3.6.1   Weight Functions

We used polynomial weight functions to precisely tune the decay of the STFT in the modulation space norm (3.6). In fact, more general families of weights can be taken into account to provide a more precise control. We collect below some basic definitions and facts, possibly under slightly more restrictive assumptions than usual. We refer the reader to [123] for generalizations and further details on the role of weights in Gabor analysis.

Definition 3.6.1

(i) We say that $v \colon \mathbb{R}^d \to (0, +\infty)$ is a weight function if it is a continuous and symmetric function in each coordinate:

$$
v(\pm x_1, \ldots, \pm x_d) = v(x_1, \ldots, x_d), \quad x \in \mathbb{R}^d.
$$

(ii) v is called *submultiplicative* if

$$
v(x + y) \leq v(x)v(y), \quad x, y \in \mathbb{R}^d.
$$

(iii) We say that a weight function m is moderate with respect to a given submultiplicative weight v (in short: v-*moderate*) if

$$
m(x + y) \lesssim v(x)m(y), \quad x, y \in \mathbb{R}^d.
$$

In general, we say that m is moderate if it is v-moderate for some submultiplicative weight v.

(iv) A weight v satisfies the *Gelfand-Raikov-Shilov (GRS) condition* if

$$\lim_{n\to\infty} v(nx)^{1/n} = 1, \quad x \in \mathbb{R}^d.$$

A standard family of weights is

$$m_{a,b,s,t}(x) := e^{a|x|^b}(1+|x|)^s \log^t(e+|x|), \quad a,b,s,t \in \mathbb{R}.$$

Note that tuning the parameters allows us to control polynomial, (sub)logarithmic and (sub)exponential rates of decay/growth. In particular,

 (i) if $a, s, t \geq 0$ and $0 \leq b \leq 1$ then $m_{a,b,s,t}$ is submultiplicative;
 (ii) if $a, s, t \in \mathbb{R}$ and $|b| \leq 1$ then $m_{a,b,s,t}$ is moderate;
(iii) if $a, s, t \geq 0$ and $0 \leq b < 1$ then $m_{a,b,s,t}$ satisfies the GRS condition.

Moreover, a convenient characterization of the GRS property is as follows: if v is a submultiplicative weight, then v satisfies the GRS condition if and only if $v(x) \lesssim e^{\epsilon|x|}$ for every $\epsilon > 0$ [100].

Some elementary properties of weight functions are collected below.

Lemma 3.6.2 *Let v be a submultiplicative weight and m be a v-moderate weight on $\mathbb{R}^d$.*

 (i) *The weight $1/v$ is v-moderate.*
 (ii) *v grows at most exponentially, namely there exist $C > 0$ and $a \geq 0$ such that $v(x) \leq Ce^{a|x|}, x \in \mathbb{R}^d$.*
(iii) *For any $x, y \in \mathbb{R}^d$,*

$$\frac{m(x)}{v(y)} \lesssim m(x-y) \lesssim m(x)v(y).$$

In particular, $1/v \lesssim m \lesssim v$ and $1/v \lesssim 1/m \lesssim v$.
 (iv) *If $s \geq 0$ then v_s is submultiplicative. Moreover, if $0 \leq |r| \leq s$, then both v_s and v_{-s} are v_r-moderate.*

In order to avoid the technical difficulties related with the exponential growth of general submultiplicative weights, in the rest of the monograph **all** the weight functions are always assumed to grow at most polynomially.

Let us denote by $\mathcal{M}_v(\mathbb{R}^{2d})$ the space of all weight functions on $\mathbb{R}^{2d}$ which are moderate with respect to an *admissible* weight function v, that is a submultiplicative weight of temperate growth (that is, $v \lesssim v_s$ for some $s \in \mathbb{R}$); the same applies when we say that m is a moderate weight. Most of the properties that have been stated above in the case of polynomial weights, for instance Proposition 3.2.3, still hold in the case where $m \in \mathcal{M}_v(\mathbb{R}^{2d})$—possibly in a more general form. The interested reader may find more precise statements and estimates in the monograph [121].

3.6.2   The Cohen Class of Time-Frequency Representations

It is well known that not all properties which are desired from a time-frequency
representation are compatible. For instance, the Wigner transform is real-valued,
but it may take negative values; this is a serious obstruction to the interpretation
of the Wigner transform as a probability distribution or as an energy density of
a signal. A key result in the problem of the positivity of the Wigner distribution is
Hudson's theorem, saying that generalized Gaussian functions have positive Wigner
transforms [150].

Theorem 3.6.3 (Hudson) *Let* $f, g \in L^2(\mathbb{R}^d)$. *We have* $W(f, g)(z) \geq 0$ *for all*
$z \in \mathbb{R}^{2d}$ *if and only if* $f = cg$ *for some* $c \geq 0$ *and* g *is a generalized Gaussian*
function, namely $g(t) = e^{Q(t)}$ *where* $Q: \mathbb{R}^d \to \mathbb{C}$ *is a quadratic polynomial such*
that $\mathrm{Re}\, Q(x) \to +\infty$ *as* $|x| \to \infty$.

The question of the zeros of the Wigner distribution is a highly non-trivial
problem which requires the contribution of several branches of analysis, see the
recent paper [127].

In order to obtain time-frequency representations that are positive for all
functions but still retaining the nice properties of the Wigner distribution (marginal
densities, orthogonality relations, etc.), one is lead to take local averages of the
Wigner transform in the hope of taming sign oscillations. This is usually done by
convolving Wf with a suitable kernel θ and such a procedure yields a general class
of quadratic time-frequency representations, which is called *Cohen's class* after L.
Cohen [43].

Time-frequency representations in Cohen's class are parametrized by a kernel
$\theta \in \mathcal{S}'(\mathbb{R}^{2d})$, in the sense of the following definition:

$$Q_\theta(f, g) := W(f, g) * \theta, \qquad f, g \in \mathcal{S}(\mathbb{R}^d). \tag{3.12}$$

The Cohen class provides a unifying framework for the study of many time-
frequency representations currently used in signal processing [44, 45, 137, 138] and
the correspondence between properties of θ and Q_θ is well understood [137]. For
instance, an important variation of the Wigner transform is given by the family of
τ-*Wigner transforms* [23, 24], which are defined in terms of a real parameter τ by

$$W_\tau(f, g)(x, \xi) = \int_{\mathbb{R}^d} e^{-2\pi i y \cdot \xi} f(x + \tau y)\overline{g(x - (1 - \tau)y)}\, dy, \quad f, g \in \mathcal{S}(\mathbb{R}^d).$$

Such distributions belong to Cohen's class, their kernel $\theta_\tau \in \mathcal{S}'(\mathbb{R}^{2d})$ being given
on the spectral side by

$$\widehat{\theta_\tau}(x, \xi) = e^{-2\pi i (\tau - 1/2) x \cdot \xi}, \quad (x, \xi) \in \mathbb{R}^{2d}.$$

It is quite tempting to stress the resistance of the Wigner distribution under more general perturbations in the Cohen class, still with some loose control on the perturbation. A general linear perturbation amounts to replace the scalar parameter τ with a matrix parameter $T \in \mathbb{R}^{d \times d}$, which leads to the family of *matrix-Wigner distributions*

$$W_T(f, g)(x, \xi) = \int_{\mathbb{R}^d} e^{-2\pi i y \cdot \xi} f(x + Ty)\overline{g(x - (I - T)y)}dy. \qquad (3.13)$$

These are members of the Cohen class in (3.12) with a kernel θ_T given by

$$\theta_T = \mathcal{F}^{-1}\Theta_T \in \mathcal{S}'(\mathbb{R}^{2d}), \qquad \Theta_T(u, v) = e^{-2\pi i \xi \cdot (T - I/2)\eta}.$$

An even more general definition in the spirit of (3.13) uses an arbitrary linear mapping of the pair $(x, y) \in \mathbb{R}^{2d}$. Let $A = \begin{bmatrix} A_{11} & A_{12} \\ A_{21} & A_{22} \end{bmatrix} \in \mathbb{R}^{2d \times 2d}$ be an invertible, real-valued $2d \times 2d$-matrix. We define the bilinear time-frequency transform $\mathcal{B}_A$ of two functions f, g by

$$\mathcal{B}_A(f, g)(x, \xi) = \int_{\mathbb{R}^d} e^{-2\pi i y \cdot \xi} f(A_{11}x + A_{12}y)\overline{g(A_{21}x + A_{22}y)}dy.$$

Note that this general framework also encompasses the Gabor transform (1.2). Clearly, W_T in (3.13) is a special case by choosing

$$A = A_T = \begin{bmatrix} I & T \\ I & -(I - T) \end{bmatrix}.$$

The interested reader may consult [11, 12, 22, 51, 65, 66, 219] for results on τ/T-Wigner distributions and the associated pseudodifferential calculi. In this connection, see also Sect. 4.5.2 below.

3.6.3 Kato-Sobolev Spaces

For $\kappa \in \mathbb{N}$ we consider the *Kato-Sobolev space* $H_{ul}^\kappa(\mathbb{R}^d)$ (also known as *uniformly local Sobolev space*) of functions f in $(L^1(\mathbb{R}^d))_{loc}$ satisfying

$$\|f\|_{H_{ul}^\kappa(\mathbb{R}^d)} := \sup_B \|f\|_{H^\kappa(B)} = \sup_B \sup_{|\alpha| \leq \kappa} \|\partial^\alpha f\|_{L^2(B)} < \infty$$

where the supremum is taken over all open balls $B \subset \mathbb{R}^d$ of radius 1 and the derivatives are understood in the sense of distributions. In fact, considering balls of any fixed radius > 0 yields equivalent norms.

In general, for every $r \in \mathbb{R}$ one defines the spaces $H_{\mathrm{ul}}^r(\mathbb{R}^d)$ of temperate distributions $f \in \mathcal{S}'(\mathbb{R}^d)$ such that

$$\|f\|_{H_{\mathrm{ul}}^r(\mathbb{R}^d)} := \sup_{y \in \mathbb{R}^d} \|\chi(\cdot - y)f\|_{H^r(\mathbb{R}^d)} < \infty,$$

where $\chi \in C_c^\infty(\mathbb{R}^d) \setminus \{0\}$. We refer to Sect. 2.4 for the definition of the norm of $H^r(\mathbb{R}^d)$. It is easy to realize that different choices of χ give rise to equivalent norms, and that the previous definitions do coincide when r is a non-negative integer. Moreover if $r > d/2$, $H_{\mathrm{ul}}^r(\mathbb{R}^d) \subset C(\mathbb{R}^d) \cap L^\infty(\mathbb{R}^d)$ is a Banach algebra [7, 27, 156].

We will use (in the proof of Proposition 8.5.1) the following scaling property.

Proposition 3.6.4 *Let $\kappa \in \mathbb{N}$ and assume $a \in H_{\mathrm{ul}}^{\kappa+[d/2]+1}(\mathbb{R}^{2d})$. Then, the function $a_t(x, y) = a(x, ty)$, $0 \le t \le 1$, $x, y \in \mathbb{R}^d$, belongs to $H_{\mathrm{ul}}^\kappa(\mathbb{R}^{2d})$ and there is a constant $C > 0$ independent of t and a such that*

$$\|a_t\|_{H_{\mathrm{ul}}^\kappa(\mathbb{R}^{2d})} \le C \|a\|_{H_{\mathrm{ul}}^{\kappa+[d/2]+1}(\mathbb{R}^{2d})},$$

where $[d/2]$ stands for the integer part of $d/2$.

Proof We use the easily verified fact that an equivalent norm in $H_{\mathrm{ul}}^\kappa(\mathbb{R}^{2d})$ is given by

$$\|f\|_{H_{\mathrm{ul}}^\kappa(\mathbb{R}^{2d})} = \sup_{B, B'} \sup_{|\alpha|+|\beta| \le \kappa} \|\partial_x^\alpha \partial_y^\beta f\|_{L^2(B \times B')}$$

where B, B' are balls of radius 1 in $\mathbb{R}^d$.

Let $|\alpha| + |\beta| \le \kappa$ and B be an open ball of radius 1 in $\mathbb{R}_y^d$. We have

$$\|\partial_x^\alpha \partial_y^\beta a_t(x, \cdot)\|_{L^2(B)} = t^{|\beta|-d/2} \|\partial_x^\alpha \partial_y^\beta a(x, \cdot)\|_{L^2(\tilde{B})}$$

where $\tilde{B} = \{ty : y \in B\}$ is a ball of radius t. Now, if $|\beta| \ge d/2$ we have $t^{|\beta|-d/2} \le 1$, so that the desired conclusion follows easily by taking the norm in $L^2(B')$ of this expression, where $B' \subset \mathbb{R}_x^d$ is any ball of radius 1.

If instead $|\beta| < d/2$, hence $|\beta| < m := [d/2] + 1$, then we set $1/p = |\beta|/(2m)$ and continue the above estimate using Hölder's inequality:

$$\|\partial_x^\alpha \partial_y^\beta a_t(x, \cdot)\|_{L^2(B)} \le C t^{|\beta|-d/2+d(1/2-1/p)} \|\partial_x^\alpha \partial_y^\beta a(x, \cdot)\|_{L^p(\tilde{B})}$$

$$= C t^{|\beta|(1-d/(2m))} \|\partial_x^\alpha \partial_y^\beta a(x, \cdot)\|_{L^p(\tilde{B})}$$

$$\le C t^{|\beta|(1-d/(2m))} \|\partial_x^\alpha \partial_y^\beta a(x, \cdot)\|_{L^p(B'')}$$

where $B'' \supseteq \tilde{B}$ is the ball of radius 1 with the same center as $\tilde{B}$. As a consequence of the Gagliardo-Nirenberg-Sobolev inequality (2.2) the latter expression is dominated by

$$C' t^{|\beta|(1-d/(2m))} \|\partial_x^\alpha a(x, \cdot)\|_{L^\infty(B'')}^{1-|\beta|/m} \|\partial_x^\alpha a(x, \cdot)\|_{H^m(B'')}^{|\beta|/m}$$

which in turn is dominated (since $m > d/2$, and therefore $t^{|\beta|(1-d/(2m))} \leq 1$ and $H^m(B'') \hookrightarrow L^\infty(B'')$) by

$$C'' \|\partial_x^\alpha a(x, \cdot)\|_{H^m(B'')},$$

and one concludes as above. $\qquad\square$

Remark 3.6.5 We stress that a loss of $d/2$ derivatives in Proposition 3.6.4 is unavoidable. Indeed, suppose that for $|\alpha| = \kappa$ and some $r \geq 0$ the following estimate holds:

$$\|\partial_x^\alpha a_t\|_{L^2(B \times B)} \leq C \|a\|_{H^{\kappa+r}_{ul}(\mathbb{R}^{2d})},$$

where $B = \{x \in \mathbb{R}^d : |x| < 1\}$. We test this estimate with $a(x, y) = a^{(1)}(\lambda x) a^{(2)}(\lambda y)$, $\lambda = 1/t \geq 1$, where $a^{(1)}(x)$, $a^{(2)}(y)$ are in $C_c^\infty(B) \setminus \{0\}$. We obtain

$$\|\partial^\alpha (a^{(1)}(\lambda \cdot))\|_{L^2(B)} \|a^{(2)}\|_{L^2(B)} \leq C' \|a^{(1)}(\lambda \cdot) a^{(2)}(\lambda \cdot)\|_{H^{\kappa+r}(\mathbb{R}^{2d})},$$

hence as $\lambda \to +\infty$ we infer $\lambda^{\kappa-d/2} \leq C'' \lambda^{\kappa+r-d}$, which implies $r \geq d/2$.

We are particularly interested in the issue of composition for Sobolev mappings.

Proposition 3.6.6 *Let $g \colon \mathbb{R}^d \to \mathbb{R}^d$ be a globally bi-Lipschitz map, i.e. satisfying*

$$C_0^{-1}|x - y| \leq |g(x) - g(y)| \leq C_0|x - y|, \quad x, y \in \mathbb{R}^d \tag{3.14}$$

for some constant $C_0 > 0$. Let Dg denote the Jacobian matrix of g.

Let $\kappa \in \mathbb{N}$. If $f \in H^\kappa_{ul}(\mathbb{R}^d) \cap L^\infty(\mathbb{R}^d)$ and $Dg \in H^\kappa_{ul}(\mathbb{R}^d; \mathbb{R}^{d \times d})$ then $f \circ g \in H^\kappa_{ul}(\mathbb{R}^d) \cap L^\infty(\mathbb{R}^d)$.

Proof The result is a variation on the theme of composition formulas for Sobolev mappings, see e.g. [35, 145] and the references therein. For the benefit of the reader we provide a short proof.

First, note that for every open ball B of radius 1 we have

$$\|f \circ g\|_{L^2(B)} \leq C \|f \circ g\|_{L^\infty(B)} \leq C \|f \circ g\|_{L^\infty(\mathbb{R}^d)} \leq C \|f\|_{L^\infty(\mathbb{R}^d)}.$$

We use now the chain rule to handle the derivatives of $f \circ g$—for the justification of the chain rule at this regularity level we refer to [35]. Let $g = (g_1, \ldots, g_d)$. We can write $\partial^\alpha (f \circ g)$, $1 \leq |\alpha| \leq \kappa$, as a linear combination of terms of the form

$$\partial^\sigma f(g(x)) \, \partial^{\mu_1} g_{j_1}(x) \cdots \partial^{\mu_{|\sigma|}} g_{j_{|\sigma|}}(x)$$

where $1 \leq |\sigma| \leq |\alpha|$, $|\mu_1|, \ldots, |\mu_{|\sigma|}| \geq 1$, $\mu_1 + \ldots + \mu_{|\sigma|} = \alpha$, and $j_1, \ldots, j_{|\sigma|} \in \{1, \ldots, d\}$.

Given an open ball $B \subset \mathbb{R}^d$ of radius 1, the $L^2(B)$-norm of this expression can be bounded via Hölder's inequality by

$$\|(\partial^\sigma f) \circ g\|_{L^{p_0}(B)} \|\partial^{\mu_1} g_{j_1}\|_{L^{p_1}(B)} \cdots \|\partial^{\mu_{|\sigma|}} g_{j_{|\sigma|}}\|_{L^{p_{|\sigma|}}(B)}, \tag{3.15}$$

where we choose

$$\frac{1}{p_0} = \frac{|\sigma|}{2\kappa}, \quad \frac{1}{p_j} = \frac{|\mu_j| - 1}{2\kappa}, \quad j = 1, \ldots, |\sigma|.$$

Indeed, note that

$$\sum_{j=0}^{|\sigma|} \frac{1}{p_j} = \frac{|\sigma|}{2\kappa} + \frac{|\alpha| - |\sigma|}{2\kappa} = \frac{|\alpha|}{2\kappa} \leq \frac{1}{2}.$$

On the other hand, by the bi-Lipschitz assumption (3.14) we have, with $\tilde{B} = g(B)$,

$$\|(\partial^\sigma f) \circ g\|_{L^{p_0}(B)} \leq C \|\partial^\sigma f\|_{L^{p_0}(\tilde{B})}.$$

Using the fact that $\tilde{B} = g(B)$ can be covered by a fixed number of balls B' of radius 1 and the Gagliardo-Nirenberg-Sobolev inequality (2.2) we can go ahead with the estimate:

$$\|(\partial^\sigma f) \circ g\|_{L^{p_0}(B)} \leq C' \sup_{B'} \|\partial^\sigma f\|_{L^{p_0}(B')}$$

$$\leq C'' \sup_{B'} \|f\|_{L^\infty(B')}^{1 - |\sigma|/\kappa} \|f\|_{H^\kappa(B')}^{|\sigma|/\kappa}$$

$$\leq C'' \|f\|_{L^\infty(\mathbb{R}^d)}^{1 - |\sigma|/\kappa} \|f\|_{H^\kappa_{\mathrm{ul}}(\mathbb{R}^d)}^{|\sigma|/\kappa}.$$

As a result, we have

$$\sup_B \|(\partial^\sigma f) \circ g\|_{L^{p_0}(B)} < \infty.$$

Similar arguments apply to the other factors in (3.15). Namely, for $j = 1, \ldots, |\sigma|$ we have $|\mu_j| \geq 1$, so that using the Gagliardo-Nirenberg-Sobolev inequality one more time we obtain

$$\|\partial^{\mu_j} g\|_{L^{p_j}(B)} \leq C \|Dg\|_{L^\infty(B)}^{1-(|\mu_j|-1)/\kappa} \|Dg\|_{H^\kappa(B)}^{(|\mu_j|-1)/\kappa},$$

which is uniformly bounded with respect to B in light of the assumptions on g— note indeed that (3.14) implies $Dg \in L^\infty(\mathbb{R}^d; \mathbb{R}^{d \times d})$. □

Let us consider now an appropriate version of the inverse mapping theorem.

Proposition 3.6.7 *Let $\kappa \in \mathbb{N}$. Let $g \colon \mathbb{R}^d \to \mathbb{R}^d$ be a globally bi-Lipschitz map, i.e., satisfying (3.14), with $Dg \in H^\kappa_{ul}(\mathbb{R}^d; \mathbb{R}^{d \times d})$. Then $Dg^{-1} \in H^\kappa_{ul}(\mathbb{R}^d; \mathbb{R}^{d \times d})$ as well.*

Proof In view of the inverse mapping theorem for Sobolev mappings (as stated for instance in [35, Theorem 1.1], with $m = \kappa + 1$ and $p = 2$) we have that $Dg^{-1} \in H^\kappa(B; \mathbb{R}^{d \times d}))$ for every open ball $B \subset \mathbb{R}^d$, even if g were only locally bi-Lipschitz. In fact, an inspection of the proof of [35, Theorem 1.1] shows that, since g is globally bi-Lipschitz and Dg is in $H^\kappa_{ul}(\mathbb{R}^d; \mathbb{R}^{d \times d})$, the desired estimates are uniform with respect to the balls B (with the same fixed radius), and the claim thus follows. □

Remark 3.6.8 It also follows from the proof of Proposition 3.6.7 that we have, in fact, $\|Dg^{-1}\|_{H^\kappa_{ul}} \leq C$ for a constant C depending only on C_0, κ, d and upper bounds for $\|Dg\|_{H^\kappa_{ul}}$, where C_0 in the constant in (3.14). A similar remark applies to Proposition 3.6.6. In Chap. 8 we will apply such "uniform versions" of Propositions 3.6.6 and 3.6.7 whenever convenient, without further comments.

3.6.4 Fourier Multipliers

We provide sufficient conditions on the symbol of a Fourier multiplier in order for it to be bounded on modulation and Wiener amalgam spaces. We will see, as a consequence of general results on the boundedness of pseudodifferential operators (Theorem 4.2.2 below), that a Fourier multiplier $m(D)$ with symbol $m \in M^{\infty,1}(\mathbb{R}^d)$ is bounded on $M^{p,q}(\mathbb{R}^d)$, $1 \leq p, q \leq \infty$. In fact better results hold, as shown by the following proposition (see also [222, Section 2.4] for a further generalization).

Proposition 3.6.9 *Let $m \in W^{1,\infty}_{|r|,\delta}(\mathbb{R}^d)$ for some $r, \delta \in \mathbb{R}$. The Fourier multiplier $m(D)$ is bounded from $M^{p,q}_{r,s}(\mathbb{R}^d)$ to $M^{p,q}_{r,s+\delta}(\mathbb{R}^d)$ for any $1 \leq p, q \leq \infty$ and $s \in \mathbb{R}$, and*

$$\|m(D)f\|_{M^{p,q}_{r,s+\delta}} \lesssim \|m\|_{W^{1,\infty}_{|r|,\delta}} \|f\|_{M^{p,q}_{r,s}}, \quad f \in M^{p,q}_{r,s}(\mathbb{R}^d).$$

Proof The proof follows the pattern in [15, Lemma 8]. Choose as window a function $g \in \mathcal{S}(\mathbb{R}^d)$ that factors as $g = g_0 * g_0$ for some $g_0 \in \mathcal{S}(\mathbb{R}^d)$—consider for instance $g_0(t) = e^{-\pi t^2}$. It is then easy to prove that $M_\xi g^* = M_\xi g_0^* * M_\xi g_0^*$. Thanks to (3.1), Propositions 2.4.1 and 3.3.3, and the associativity of convolutions we have

$$
\begin{aligned}
\|m(D)f\|_{M_{r,s+\delta}^{p,q}} &= \left(\int_{\mathbb{R}^d} \|m(D)f * M_\xi g^*\|_{L_r^p}^q v_{s+\delta}(\xi)d\xi \right)^{1/q} \\
&= \left(\int_{\mathbb{R}^d} \|\check{m} * f * M_\xi g_0^* * M_\xi g_0^*\|_{L_r^p}^q v_{s+\delta}(\xi)d\xi \right)^{1/q} \\
&= \left(\int_{\mathbb{R}^d} \|(\check{m} * M_\xi g_0^*) * (f * M_\xi g_0^*)\|_{L_r^p}^q v_{s+\delta}(\xi)d\xi \right)^{1/q} \\
&\leq \left(\int_{\mathbb{R}^d} \|\check{m} * M_\xi g_0^*\|_{L_{|r|}^1}^q \|f * M_\xi g_0^*\|_{L_r^p}^q v_{s+\delta}(\xi)d\xi \right)^{1/q} \\
&\leq \left(\sup_{\xi \in \mathbb{R}^d} \|\check{m} * M_\xi g_0^*\|_{L_{|r|}^1} v_\delta(\xi) \right) \left(\int_{\mathbb{R}^d} \|f * M_\xi g_0^*\|_{L_r^p}^q v_s(\xi)d\xi \right)^{1/q} \\
&= \|m\|_{W_{|r|,\delta}^{1,\infty}} \|f\|_{M_{r,s}^{p,q}}.
\end{aligned}
$$

The cases where $p = \infty$ or $q = \infty$ can be handled similarly after slight modifications. ☐

A similar result holds for Fourier multipliers on Wiener amalgam spaces.

Proposition 3.6.10 *Let* $m \in M_{\delta,|s|}^{\infty,1}(\mathbb{R}^d)$ *for some* $s, \delta \in \mathbb{R}$. *The Fourier multiplier* $m(D)$ *is bounded from* $W_{r,s}^{p,q}(\mathbb{R}^d)$ *to* $W_{r+\delta,s}^{p,q}(\mathbb{R}^d)$ *for any* $1 \leq p, q \leq \infty$ *and* $r \in \mathbb{R}$, *and*

$$
\|m(D)f\|_{W_{r+\delta,s}^{p,q}} \lesssim \|m\|_{M_{\delta,|s|}^{\infty,1}} \|f\|_{W_{r,s}^{p,q}}, \quad f \in W_{r,s}^{p,q}(\mathbb{R}^d).
$$

Proof Recall that $W_{r,s}^{p,q}(\mathbb{R}^d) = W(\mathcal{F}L_r^p, L_s^q)(\mathbb{R}^d) = \mathcal{F}M_{r,s}^{p,q}(\mathbb{R}^d)$. Theorem 3.3.3 thus yields

$$
\begin{aligned}
\|m(D)f\|_{W_{r+\delta,s}^{p,q}} &= \left\| \mathcal{F}^{-1}m * f \right\|_{W_{r+\delta,s}^{p,q}} \\
&\lesssim \left\| \mathcal{F}^{-1}m \right\|_{W_{\delta,|s|}^{\infty,1}} \|f\|_{W_{r,s}^{p,q}} \\
&\lesssim \|m\|_{M_{\delta,|s|}^{\infty,1}} \|f\|_{W_{r,s}^{p,q}}.
\end{aligned}
$$

☐

3.6.5 More on the Sjöstrand Class

Proposition 3.5.5 can be generalized to classes of unbounded functions. We present the following result (cf. [222, Proposition 3.2]), because the same idea will be used in Proposition 8.4.2 below. We refer to Sect. 2.2 for the definition of the space $C^\infty_{\geq k}(\mathbb{R}^d)$.

Proposition 3.6.11 Let $f: \mathbb{R}^d \to \mathbb{C}$ be such that $\partial^\alpha f \in M^{\infty,1}(\mathbb{R}^d)$ for any $\alpha \in \mathbb{N}^d$ such that $|\alpha| = k$ for some $k \in \mathbb{N}$. Then there exist $f_1 \in C^\infty_{\geq k}(\mathbb{R}^d)$ and $f_2 \in M^{\infty,1}(\mathbb{R}^d)$ such that $f = f_1 + f_2$.

Proof Fix a smooth cut-off function $\chi \in C^\infty_c(\mathbb{R}^d)$ supported in a neighbourhood of the origin and such that $\chi = 1$ near zero, then consider the Fourier multiplier $\chi(D)$ with symbol χ. Set $f_1 = \chi(D)f$ and $f_2 = (I - \chi(D))f$. Clearly $f = f_1 + f_2$ and we argue that f_1 and f_2 satisfy the claimed properties.

Indeed, $f_1 \in C^\infty(\mathbb{R}^d)$ and for any $\alpha \in \mathbb{N}^d$, $|\alpha| = k$, we have

$$\partial^\alpha f_1 = \partial^\alpha(\chi(D)f) = \chi(D)(\partial^\alpha f) \in M^{\infty,1}(\mathbb{R}^d),$$

since $\partial^\alpha \chi(D)$ is a Fourier multiplier with symbol $(2\pi i\xi)^\alpha \chi(\xi) \in C^\infty_c(\mathbb{R}^d)$, hence $\partial^\alpha \chi(D) = \chi(D)\partial^\alpha$ and $\chi(D)$ is continuous on $M^{\infty,1}(\mathbb{R}^d)$ by Proposition 3.6.9. Furthermore, similar arguments imply that for any $\alpha \in \mathbb{N}^d$, $|\alpha| \geq k$,

$$\partial^\alpha f_1 = \partial^{\alpha-\beta}\partial^\beta(\chi(D)f) = (\partial^{\alpha-\beta}\chi(D))(\partial^\beta f) \in M^{\infty,1}(\mathbb{R}^d).$$

where $\beta \in \mathbb{N}^d$ satisfies $|\beta| = k$. In order to prove the claim for f_2 consider the finite smooth partition of unity $\{\varphi_j\}_{j=1}^N$ of the unit sphere $S^{d-1} \subset \mathbb{R}^d$ subordinated to the open cover given by the family $\{U_j\}_{j=1}^d$ of sets $U_j = \{x \in S^{d-1} : x_j \neq 0\}$. We also extend each function f_j on $\mathbb{R}^d \setminus \{0\}$ by zero-degree homogeneity, namely

$$\sum_{j=1}^d \varphi_j(x) = 1, \qquad \varphi_j(\alpha x) = \varphi_j(x), \qquad \forall x \in S^{d-1}, \ \alpha > 0.$$

This procedure gives a finite partition of unity $\{\varphi_j\}_{k=1}^d$ on $\mathbb{R}^d \setminus \{0\}$. Then

$$f_2(x) = \int_{\mathbb{R}^d} e^{2\pi ix\cdot\xi}(1 - \chi(\xi))\hat{f}(\xi)d\xi$$

$$= \sum_{j=1}^d \left[\int_{\mathbb{R}^d} e^{2\pi ix\cdot\xi} \left(\frac{1 - \chi(\xi)}{(2\pi i\xi_j)^k}\varphi_j(\xi) \right) \widehat{\partial_j^k f}(\xi)d\xi \right]$$

$$= \sum_{j=1}^d \tilde{\chi}_j(D)(\partial_j^k f)(x)$$

and thus $f_2 \in M^{\infty,1}(\mathbb{R}^d)$ since each $\widetilde{\chi}_j(D)$ is a Fourier multiplier with symbol $(1 - \chi(\xi))\varphi_j(\xi)/(2\pi i \xi_j)^k \in C_{\geq 0}^{\infty}(\mathbb{R}^d) \subset M^{\infty,1}(\mathbb{R}^d)$, hence bounded on $M^{\infty,1}(\mathbb{R}^d)$ by Proposition 3.6.9.   □

Narrow Convergence Convergence in $M^{\infty,1}$ norm happens to be a quite strong condition for the purposes of several applications. As a basic example, the fact that C_c^{∞} is not dense in $M^{\infty,1}$ with the norm topology [209] is a substantial obstruction to the standard approximation arguments, which can be partially circumvented by restriction to suitable subspaces (e.g., the completion of C_c^{∞} under the $M^{\infty,1}$-norm). Another way to cope with this problem consists in weakening the notion of convergence as follows [58, 209].

Definition 3.6.12 Let Ω be a subset of some Euclidean space. The map $\Omega \ni \nu \mapsto \sigma_\nu \in M^{\infty,1}(\mathbb{R}^d)$ is said to be continuous for the narrow convergence if:

 (i) it is a continuous map in $\mathcal{S}'(\mathbb{R}^d)$ (weakly), and
 (ii) there exists a function $h \in L^1(\mathbb{R}^d)$ such that for some (hence any) nonzero window $g \in \mathcal{S}(\mathbb{R}^d)$ one has $\sup_{x \in \mathbb{R}^d} |V_g \sigma_\nu(x, \xi)| \leq h(\xi)$ for any $\nu \in \Omega$ and a.e. $\xi \in \mathbb{R}^d$.

It turns out that $\mathcal{S}(\mathbb{R}^d)$ is dense in $M^{\infty,1}$ with respect to the narrow convergence [209]. This notion of convergence will be used in a result in Sect. 9.7 below.

3.6.6 Boundedness of Time-Frequency Transforms on Modulation Spaces

For the sake of completeness we recall necessary and sufficient conditions for the boundedness of the Wigner transform on modulation spaces [48, Theorem 1.1].

Theorem 3.6.13 Let $p_i, q_i, p, q \in [1, \infty]$, $s \in \mathbb{R}$, be such that

$$p_i, q_i \leq q, \quad i = 1, 2 \tag{3.16}$$

and

$$\frac{1}{p_1} + \frac{1}{p_2} \geq \frac{1}{p} + \frac{1}{q}, \quad \frac{1}{q_1} + \frac{1}{q_2} \geq \frac{1}{p} + \frac{1}{q}. \tag{3.17}$$

If $f_1 \in M_{v_{|s|}}^{p_1,q_1}(\mathbb{R}^d)$ and $f_2 \in M_{v_s}^{p_2,q_2}(\mathbb{R}^d)$ then $W(f_1, f_2) \in M_{0,s}^{p,q}(\mathbb{R}^{2d})$, with

$$\|W(f_1, f_2)\|_{M_{0,s}^{p,q}} \lesssim \|f_1\|_{M_{v_{|s|}}^{p_1,q_1}} \|f_2\|_{M_{v_s}^{p_2,q_2}}.$$

Conversely, assume that there exists a constant $C > 0$ such that

$$\|W(f_1, f_2)\|_{M^{p,q}} \leq C \|f_1\|_{M^{p_1, q_1}} \|f_2\|_{M^{p_2, q_2}}, \quad \forall f_1, f_2 \in \mathcal{S}(\mathbb{R}^d).$$

Then (3.16) and (3.17) must hold.

Note that the previous result extends to the ambiguity function and the Gabor transform, in view of the following equivalence (cf. [48, Lemma 4.1]):

$$\|W(f_1, f_2)\|_{M_{0,s}^{p,q}} = \|\mathrm{Amb}(f_1, f_2)\|_{W_{0,s}^{p,q}} \asymp \|V_{f_2} f_1\|_{W_{0,s}^{p,q}}.$$

3.6.7 Gabor Frames

We already stressed that the STFT $V_g f$ may be heuristically interpreted as a "continuous expansion" of f in terms of highly localized wave packets $\{\pi(z)g : z = (x, \xi) \in \mathbb{R}^{2d}\}$. The theory of Gabor frames may be used to give a precise meaning to this suggestion in terms of discrete samples and expansions. Although we do not resort to discrete decompositions in this book, we sketch the basics of the theory.

To be precise, given a non-zero window function $g \in L^2(\mathbb{R}^d)$ and a *full-rank lattice* $\Lambda \subset \mathbb{R}^{2d}$ (namely, a countable and discrete additive subgroup of $\mathbb{R}^{2d}$ with compact quotient group $\mathbb{R}^{2d}/\Lambda$), the *Gabor system* $\mathcal{G}(g, \Lambda)$ is the collection of time-frequency shifts of g along Λ, namely

$$\mathcal{G}(g, \Lambda) = \{\pi(z)g : z \in \Lambda\}.$$

Standard examples of lattices are $\Lambda = M\mathbb{Z}^{2d}$ where $M \in \mathrm{GL}(2d, \mathbb{R})$, in particular separable lattices such as

$$\Lambda = \alpha\mathbb{Z} \times \beta\mathbb{Z} = \{(\alpha k, \beta n) : k, n \in \mathbb{Z}\},$$

for suitable lattice parameters $\alpha, \beta > 0$; we write $\mathcal{G}(g, \alpha, \beta)$ for the corresponding Gabor system.

Recall that a *frame* for $L^2(\mathbb{R}^d)$ is a sequence $\{\phi_j\}_{j \in J} \subset L^2(\mathbb{R}^d)$, J being a countable index set, such that for all $f \in L^2(\mathbb{R}^d)$

$$A\|f\|_{L^2}^2 \leq \sum_{j \in J} |\langle f, \phi_j \rangle|^2 \leq B\|f\|_{L^2}^2,$$

for some positive constants $A, B > 0$ (frame bounds). We say that a frame is tight if $A = B$ and Parseval if $A = B = 1$. The reader may consult the monographs [42, 134] and the references therein.

One may naturally define some bounded linear operators related to a given frame:

- The *analysis operator* $C \colon L^2(\mathbb{R}^d) \to \ell^2(J)$, $Cf = \{\langle f, \phi_j \rangle\}_{j \in J}$.
- The *synthesis operator* $D \colon \ell^2(J) \to L^2(\mathbb{R}^d)$, $Da = \sum_{j \in J} a_j \phi_j$.
- The *frame operator* $S \colon L^2(\mathbb{R}^d) \to L^2(\mathbb{R}^d)$, $Sf = DCf = \sum_{j \in J} \langle f, \phi_j \rangle \phi_j$.

If a Gabor system $\mathcal{G}(g, \Lambda)$ is a frame for $L^2(\mathbb{R}^d)$ it is called *Gabor frame*. Notice that the Gabor frame operator reads

$$Sf = \sum_{\lambda \in \Lambda} V_g f(\lambda) \pi(\lambda) g,$$

and is a positive, bounded invertible operator on $L^2(\mathbb{R}^d)$. A remarkable result of frame theory is that a function can be reconstructed from its Gabor coefficients, in the sense that

$$f = \sum_{\lambda \in \Lambda} V_g f(\lambda) \pi(\lambda) \gamma, \tag{3.18}$$

where $\gamma = S^{-1} g$ is the *canonical dual window* and the sum is unconditionally convergent in L^2.

Discrete Gabor analysis can be extended to modulation spaces under suitable assumptions (cf. [121, Corollary 12.2.6 and Corollary 12.2.8]); for instance, if $g \in M^1(\mathbb{R}^d)$ and $\mathcal{G}(g, \alpha, \beta)$ is a tight frame for $L^2(\mathbb{R}^d)$ then for any $1 \le p \le \infty$,

$$\|f\|_{M_m^{p,q}} = \frac{1}{A} \left(\sum_{n \in \mathbb{Z}^d} \left(\sum_{k \in \mathbb{Z}^d} |V_g f(\alpha k, \beta n)|^p m(\alpha k, \beta n)^p| \right)^{q/p} \right)^{1/q},$$

and the Gabor expansion (3.18) holds with $\gamma = g/A$ and unconditional convergence if $1 \le p < \infty$ (weak*-convergence if $p = \infty$).

In conclusion, we mention that a famous open problem in Gabor analysis is the so-called *HRT conjecture* on the linear independence of finitely many time-frequency shifts of a non-trivial $g \in L^2(\mathbb{R}^d)$. The question was first posed in 1996 [136] and can be formulated as follows.

Conjecture Given $g \in L^2(\mathbb{R}^d) \setminus \{0\}$ and a set Λ of finitely many distinct points $z_1, \ldots, z_N \in \mathbb{R}^{2d}$, then $\mathcal{G}(g, \Lambda) = \{\pi(z_k) g\}_{k=1}^N$ is a linearly independent set of functions in $L^2(\mathbb{R}^d)$.

While this problem is still open in general at the time of writing, the conjecture has been proved for certain classes of functions or for special arrangements of points. The interested reader may want to consult the papers [125, 133, 135, 185] and the references therein.

Chapter 4
The Gabor Analysis of Operators

In this chapter we expand the discussion on the analysis of operators by means of Gabor wave packets initiated in Sect. 1.2. Most of the results in this part can be found scattered in the monographs [16, 50, 71, 106, 121], but here we focus on the topics that are relevant for our purposes. This implies that we avoid unnecessary technicalities in order to provide a practical and largely self-contained path, with simplified arguments whenever possible.

4.1 The General Program

We already anticipated that the inversion formula for the short-time Fourier transform in (3.3) enables an efficient phase-space analysis of operators. Consider a continuous linear operator $A: S(\mathbb{R}^d) \to S'(\mathbb{R}^d)$ and $g, \gamma \in S(\mathbb{R}^d)$ with $\|g\|_{L^2} = \|\gamma\|_{L^2} = 1$. Therefore, we have

$$A = V_\gamma^* V_\gamma A V_g^* V_g = V_\gamma^* \widetilde{A} V_g,$$

where $\widetilde{A} := V_\gamma A V_g^*$ is an integral operator in $\mathbb{R}^{2d}$ with continuous integral kernel given by the *Gabor matrix* K_A, that is

$$\widetilde{A}F(w) = \int_{\mathbb{R}^{2d}} K_A(w, z) F(z) dz, \quad K_A(w, z) = \langle A\pi(z)g, \pi(w)\gamma \rangle, \quad w, z \in \mathbb{R}^{2d}.$$
(4.1)

This decomposition will be of primary concern for the time frequency-analysis of operators. We are now ready to consider some relevant classes of operators and derive fundamental boundedness estimates that will be repeatedly used to prove the main results.

© The Author(s), under exclusive license to Springer Nature Switzerland AG 2022   73
F. Nicola, S. I. Trapasso, *Wave Packet Analysis of Feynman Path Integrals*,
Lecture Notes in Mathematics 2305, https://doi.org/10.1007/978-3-031-06186-8_4

4.2   The Weyl Quantization

In this section we provide some essential notions on the Weyl transform and some results on the phase-space analysis of such operators.

The standard definition in PDEs of the Weyl transform $\sigma^{w} = \mathrm{op_w}(\sigma)$ of the symbol $\sigma : \mathbb{R}^{2d} \to \mathbb{C}$ is

$$\sigma^{w} f(x) := \int_{\mathbb{R}^{2d}} e^{2\pi i (x-y)\cdot\xi} \sigma\left(\frac{x+y}{2}, \xi\right) f(y)\,dy\,d\xi.$$

The meaning of this expression clearly depends on the function spaces that are considered for the choice of the symbol σ and the function f. As a matter of fact, virtually all the classical symbol classes introduced throughout the history of pseudodifferential calculus are defined by means of decay or smoothness conditions, as for instance the Hörmander classes $S^{m}_{\rho,\delta}(\mathbb{R}^{2d})$ in [141]. In general, one can interpret the above operator in weak sense on $\mathcal{S}(\mathbb{R}^{d})$, and then it makes sense for every $\sigma \in \mathcal{S}'(\mathbb{R}^{2d})$ and defines a linear continuous operator $\mathcal{S}(\mathbb{R}^{d}) \to \mathcal{S}'(\mathbb{R}^{d})$ with distribution kernel (formally written as an integral)

$$k_{\sigma}(x, y) := \int_{\mathbb{R}^{d}} e^{2\pi i (x-y)\cdot\xi} \sigma\left(\frac{x+y}{2}, \xi\right) d\xi.$$

Morevover the map $\sigma \mapsto k_{\sigma}$ is easily seen to be an automorphism of $\mathcal{S}'(\mathbb{R}^{2d})$.

We embrace below the perspective of time-frequency analysis [119] and define equivalently the Weyl quantization of a rough symbol $\sigma \in \mathcal{S}'(\mathbb{R}^{2d})$ via duality as follows:

$$\sigma^{w} : \mathcal{S}(\mathbb{R}^{d}) \to \mathcal{S}'(\mathbb{R}^{d}), \qquad \langle \sigma^{w} f, g \rangle := \langle \sigma, W(g, f) \rangle, \qquad \forall f, g \in \mathcal{S}(\mathbb{R}^{d}).$$

It is clear from this formula, using the property $W(f, g) = \overline{W(g, f)}$, that if the distribution σ is real ($\sigma = \overline{\sigma}$) then σ^{w} is formally self-adjoint, namely

$$\langle \sigma^{w} f, g \rangle = \langle f, \sigma^{w} g \rangle, \qquad f, g \in \mathcal{S}(\mathbb{R}^{d}).$$

By the same formula, taking the symplectic Fourier transform of both σ and $W(f, g)$ and using

$$\overline{\mathrm{Amb}\,(g, f)}(x, \xi) = \langle M_{\xi} T_{x/2} f, T_{-x/2} g \rangle = e^{-\pi i x\cdot\xi} \langle M_{\xi} T_{x} f, g \rangle,$$

one obtains yet another formula for σ^{w}—the so-called spreading representation:

$$\sigma^{w} f := \int_{\mathbb{R}^{2d}} \hat{\sigma}(\xi, -x) e^{-\pi i x\cdot\xi} \pi(x, \xi) f \, dx\, d\xi, \tag{4.2}$$

again interpreted in weak sense.

Remark 4.2.1 The multiplication by a function $V(x)$ is a special example of a Weyl operator with symbol

$$\sigma_V(x, \xi) = V(x) = (V \otimes 1)(x, \xi), \quad (x, \xi) \in \mathbb{R}^{2d}.$$

It is not difficult to prove that the correspondence $V \mapsto \sigma_V$ is continuous from $M_{0,s}^{\infty,q}(\mathbb{R}^d)$ to $M_{0,s}^{\infty,q}(\mathbb{R}^{2d})$ for any $1 \le q \le \infty$ and $s \in \mathbb{R}$, cf. (3.7). This identification shall be implicitly assumed whenever needed below; by a slight abuse of notation, we will write V also for σ_V^w for the sake of legibility.

One may similarly prove that a Fourier multiplier with symbol $m(\xi)$ is a Weyl operator with symbol $\tilde{\sigma}_m(x, \xi) = m(\xi) = (1 \otimes m)(x, \xi)$.

Modulation spaces can be used both as symbol classes as well a general framework for pseudodifferential operator theory. For instance, we mention that symbols in the Sjöstrand class yield Weyl operators which are bounded on any modulation space [121, Theorem 14.5.2]. This can be interpreted as a generalization of the classical result by Caldéron and Vaillancourt [33], since the space $C_b^k(\mathbb{R}^{2d})$ of k-times continuously differentiable functions with bounded derivatives up to k-th order is embedded in $M^{\infty,1}(\mathbb{R}^{2d})$ for $k > 2d$ by Proposition 3.5.1.

Theorem 4.2.2 *If $\sigma \in M^{\infty,1}(\mathbb{R}^{2d})$ then σ^w is bounded on $M^{p,q}(\mathbb{R}^d)$ for all $1 \le p, q \le \infty$, with*

$$\|\sigma^w\|_{M^{p,q} \to M^{p,q}} \lesssim \|\sigma\|_{M^{\infty,1}}.$$

The proof of this result and the subsequent theorem are given below.

The symbolic calculus relies on the composition of Weyl transforms, which provides a bilinear form on symbols known as the *Weyl* (or *twisted*) *product*:

$$\sigma^w \circ \rho^w = (\sigma \# \rho)^w, \quad \sigma \# \rho = \mathcal{F}^{-1}(\hat{\sigma} \natural \hat{\rho}),$$

where the *twisted convolution* [121] of $\hat{\sigma}$ and $\hat{\rho}$ is (formally) defined by

$$(\hat{\sigma} \natural \hat{\rho})(x, \xi) := \int_{\mathbb{R}^{2d}} e^{\pi i (x, \xi) \cdot J(y, \eta)} \hat{\sigma}(y, \eta) \hat{\rho}(x - y, \xi - \eta) \, dy d\eta.$$

Although more explicit formulas for the twisted product of symbols can be derived (cf. [240]), we will not need them hereafter. Nevertheless, this is a fundamental notion in order to establish an algebra structure on symbol spaces; the problem has been studied in several papers (cf. for instance [56, 120, 140]).

Remark 4.2.3 It is a quite distinctive property of $M^{\infty,1}(\mathbb{R}^{2d})$ as well as the scale of spaces $M_{0,s}^{\infty}(\mathbb{R}^{2d})$ with $s > 2d$, to enjoy a double Banach algebra structure:

- A commutative one with respect to the pointwise multiplication as detailed in Proposition 3.5.3.
- A non-commutative one with respect to the twisted product of symbols [126, 209].

In fact, much more is true, as detailed in the following statement [126, 209].

Theorem 4.2.4 *Let X be any of the spaces $C_b^\infty(\mathbb{R}^{2d})$, $M_{0,s}^\infty(\mathbb{R}^{2d})$ with $s > 2d$ or $M^{\infty,1}(\mathbb{R}^{2d})$. The corresponding family of Weyl operators $\mathrm{op_w}(\sigma)$, $\sigma \in X$, is a Wiener subalgebra of $\mathcal{L}(L^2(\mathbb{R}^d))$ under composition, that is:*

(i) *(Boundedness) If $\sigma \in X$, then σ^w is a bounded operator on any modulation space $M^{p,q}(\mathbb{R}^d)$, $1 \le p, q \le \infty$, in particular on $L^2(\mathbb{R}^d)$.*
(ii) *(Algebra property) If $\sigma_1, \sigma_2 \in X$ then $\sigma_1^w \circ \sigma_2^w$ is a Weyl operator with symbol $\sigma_1 \# \sigma_2 \in X$.*
(iii) *(Wiener property) If $\sigma \in X$ and σ^w is invertible on $L^2(\mathbb{R}^d)$, then there exists $\rho \in X$ such that $(\sigma^w)^{-1} = \rho^w$.*

We stress that the latter algebra structure is intimately related to a property that actually characterizes pseudodifferential operators having symbols in those spaces, namely *almost diagonalization* of the Gabor matrix by time-frequency shifts. We address the reader to [54, 55, 122, 126] for further discussions on these aspects.

Theorem 4.2.5 *Fix $g, \gamma \in \mathcal{S}(\mathbb{R}^d) \setminus \{0\}$ and consider $\sigma \in \mathcal{S}'(\mathbb{R}^{2d})$.*

(i) *$\sigma \in M_{0,s}^\infty(\mathbb{R}^{2d})$ if and only if*

$$|\langle \sigma^w \pi(z)g, \pi(w)\gamma \rangle| \lesssim_s (1 + |w - z|)^{-s}, \quad z, w \in \mathbb{R}^{2d}.$$

(ii) *$\sigma \in M^{\infty,1}(\mathbb{R}^{2d})$ if and only if there exists a function $H \in L^1(\mathbb{R}^{2d})$ such that*

$$|\langle \sigma^w \pi(z)g, \pi(w)\gamma \rangle| \le H(w - z), \quad z, w \in \mathbb{R}^{2d}.$$

The controlling function H can be chosen as

$$H_0(v) = \sup_{u \in \mathbb{R}^{2d}} |V_\Phi \sigma(u, Jv)|, \quad \Phi = W(\gamma, g),$$

hence $\|H_0\|_{L^1} \asymp \|\sigma\|_{M^{\infty,1}}$.

Proof Let us set $\Phi = W(\gamma, g)$. We have $\Phi \in \mathcal{S}(\mathbb{R}^{2d})$ as a consequence of Proposition 3.1.3 (ii), hence the Gabor transform $V_\Phi \sigma$ of $\sigma \in \mathcal{S}'(\mathbb{R}^{2d})$ is well defined.

Let us prove (ii)—the proof for $\sigma \in M_{0,s}^\infty(\mathbb{R}^{2d})$, namely (i), goes along the same arguments.

Using the covariance property (3.4) we have

$$|\langle \sigma^w \pi(z)g, \pi(w)\gamma | = |\langle \sigma, W(\pi(w)\gamma, \pi(z)g) \rangle|$$

$$= |\langle \sigma, M_{J(w-z)} T_{\frac{w+z}{2}} W(\gamma, g) \rangle|$$

$$= \left| V_\Phi \sigma \left(\frac{w+z}{2}, J(w-z) \right) \right|.$$

If read backwards, this formula yields

$$|V_\Phi \sigma(u, v)| = \left| \langle \sigma^w \pi(u - \frac{1}{2} J^{-1}v)g, \pi(u + \frac{1}{2} J^{-1}v)\gamma \rangle \right|, \quad u, v \in \mathbb{R}^{2d}. \quad (4.3)$$

Hence, if $\sigma \in M^{\infty,1}(\mathbb{R}^{2d})$ then

$$|\langle \sigma^w \pi(z)g, \pi(w)\gamma \rangle| = \left| V_\Phi \sigma \left(\frac{w+z}{2}, J(w-z) \right) \right|$$

$$\leq \sup_{u \in \mathbb{R}^{2d}} |V_\Phi(u, J(w-z))|$$

$$= H_0(w - z),$$

where $H_0 \in L^1(\mathbb{R}^{2d})$ is the controlling function defined in the claim.

Conversely, assume that $\sigma \in M^{\infty,1}(\mathbb{R}^{2d})$ and that σ^w is almost diagonalized by Gabor wave packets with controlling function $H \in L^1(\mathbb{R}^{2d})$. From (4.3) we infer

$$|V_\Phi \sigma(u, v)| \leq H(J^{-1}v), \quad u, v \in \mathbb{R}^{2d},$$

hence

$$\|\sigma\|_{M^{\infty,1}} \asymp \int_{\mathbb{R}^{2d}} \sup_{u \in \mathbb{R}^{2d}} |V_\Phi(u, v)| dv$$

$$\leq \int_{\mathbb{R}^{2d}} H(J^{-1}v)dv = \|H\|_{L^1} < \infty.$$

□

As a consequence we can now prove the above mentioned results.

Proof of Theorem 4.2.4 We exploit the ideas introduced in Sect. 4.1 above in the case where $A = \sigma^w$. In accordance with (4.1), the corresponding phase-space representation $\widetilde{\sigma^w} = V_\gamma \sigma^w V_g$ is an integral operator in $\mathbb{R}^{2d}$, its kernel being the Gabor matrix of σ^w, that is

$$\widetilde{\sigma^w} F(w) = \int_{\mathbb{R}^{2d}} K(w, z) F(z) dz, \quad K(w, z) = \langle \sigma^w \pi(z)g, \pi(w)\gamma \rangle.$$

We provide the proof in the case $X = M^{\infty,1}(\mathbb{R}^{2d})$. Identical arguments lead to the same conclusion if $X = M^{\infty}_{0,s}(\mathbb{R}^{2d})$, $s > 2d$.

(Boundedness). Recall from Proposition 3.2.3 that $V_g: M^{p,q}(\mathbb{R}^d) \to L^{p,q}(\mathbb{R}^{2d})$ and $V_\gamma^*: L^{p,q}(\mathbb{R}^{2d}) \to M^{p,q}(\mathbb{R}^d)$ are bounded operators for $1 \leq p, q \leq \infty$. Therefore, in order to prove that $\sigma^w: M^{p,q}(\mathbb{R}^d) \to M^{p,q}(\mathbb{R}^d)$ continuously, it is enough to prove that $\widetilde{\sigma^w}$ is continuous on $L^{p,q}(\mathbb{R}^{2d})$. Now, if $\sigma \in M^{\infty,1}(\mathbb{R}^{2d})$ then $|K(w, z)| \leq H(w - z)$ for some controlling function $H \in L^1(\mathbb{R}^{2d})$ in view of

Theorem 4.2.5. As a result, Young's inequality implies that

$$\|\widetilde{\sigma^w} F\|_{L^{p,q}} \leq \|H * |F|\|_{L^{p,q}} \leq \|H\|_{L^1} \|F\|_{L^{p,q}},$$

hence $\widetilde{\sigma^w} \in \mathcal{L}(L^p(\mathbb{R}^{2d}))$ as claimed.

(Algebra property). Let us assume $\sigma_1, \sigma_2 \in M^{\infty,1}(\mathbb{R}^{2d})$; recall that $\sigma_1^w \sigma_2^w = \sigma^w$, where $\sigma = \sigma_1 \# \sigma_2$. Arguing as above at the level of phase space, we get

$$|\langle \sigma^w \pi(z)g, \pi(w)\gamma\rangle| = |\langle \sigma_1^w \sigma_2^w \pi(z)g, \pi(w)\gamma\rangle|$$

$$\leq \int_{\mathbb{R}^{2d}} |K_{\sigma_1^w}(w, u)||K_{\sigma_2^w}(u, z)|du$$

$$\leq \int_{\mathbb{R}^{2d}} H_1(w - u)H_2(u - z)du$$

$$= H_1 * H_2(w - z),$$

where $H_1, H_2 \in L^1(\mathbb{R}^{2d})$ are the controlling functions for σ_1, σ_2 respectively, as in Theorem 4.2.5. The characterization in Theorem 4.2.5 implies that $\sigma \in M^{\infty,1}(\mathbb{R}^{2d})$, after setting $H = H_1 * H_2 \in L^1(\mathbb{R}^{2d})$.

The proof of the Wiener property would lead us too far. We just refer to the articles [122, 126] and to Gröchenig's lecture in [124], where the multifaceted problem of spectral invariance is approached from a general point of view.   □

4.3   Metaplectic Operators

4.3.1   Notable Facts on Symplectic Matrices

We recall the definition and the main properties of symplectic matrices, cf. [71] and the references therein for further details.

The canonical symplectic matrix $J \in \mathbb{R}^{2d \times 2d}$ is

$$J = \begin{bmatrix} O & I \\ -I & O \end{bmatrix}$$

and defines the standard symplectic structures of $\mathbb{R}^{2d}$, given by the bilinear antisymmetric non-degenerate form

$$(z_1, z_2) \mapsto Jz_1 \cdot z_2, \qquad z_1, z_2 \in \mathbb{R}^{2d}.$$

Note that $J^\top = J^{-1} = -J$. An invertible matrix $S \in \mathrm{GL}(2d, \mathbb{R})$ is said to be symplectic if $S^\top JS = J$ or equivalently if it preserves the aforementioned symplectic form. In this case, the matrices $S^\top$ and S^{-1} are symplectic too; note that the product of two symplectic matrices is clearly symplectic. As a result, the set of all symplectic matrices is a (Lie) group; hence we define the real symplectic group $\mathrm{Sp}(d, \mathbb{R})$ as

$$\mathrm{Sp}(d, \mathbb{R}) = \{S \in \mathrm{GL}(2d, \mathbb{R}) : S^\top J S = J\}.$$

We list below some well-known properties of symplectic matrices.

Proposition 4.3.1

(i) *The eigenvalues of a symplectic matrix $S \in \mathrm{Sp}(d, \mathbb{R})$ occur in quadruples, meaning that if $\lambda \in \mathbb{C} \setminus \{0\}$ is an eigenvalue of S then so are $\bar{\lambda}$ and $1/\lambda$ - hence $1/\bar{\lambda}$.*

(ii) *If $S \in \mathrm{Sp}(d, \mathbb{R})$ then $\det S = 1$.*

(iii) *Let $S \in \mathrm{GL}(2d, \mathbb{R})$ have the following block structure:*

$$S = \begin{bmatrix} A & B \\ C & D \end{bmatrix}.$$

Then $S \in \mathrm{Sp}(d, \mathbb{R})$ if and only if any of the two sets of conditions are satisfied:

$$A^\top C, \ B^\top D \text{ are symmetric, and } A^\top D - C^\top B = I;$$

$$AB^\top, \ CD^\top \text{ are symmetric, and } AD^\top - CB^\top = I.$$

Moreover, in that case the inverse matrix is explicitly given by

$$S^{-1} = \begin{bmatrix} D^\top & -B^\top \\ -C^\top & A^\top \end{bmatrix}.$$

(iv) *The complex unitary group $\mathrm{U}(d, \mathbb{C})$ is isomorphic to the subgroup of symplectic rotations $\mathrm{U}(2d, \mathbb{R})$ of $\mathrm{Sp}(d, \mathbb{R})$ defined by $\mathrm{U}(2d, \mathbb{R}) = \mathrm{Sp}(d, \mathbb{R}) \cap \mathrm{O}(2d, \mathbb{R})$. In particular, the following characterization holds:*

$$\mathrm{U}(2d, \mathbb{R}) = \left\{ \begin{bmatrix} A & -B \\ B & A \end{bmatrix} \in \mathbb{R}^{2d \times 2d} : AA^\top + BB^\top = I, \ AB^\top = B^\top A \right\}.$$

Here $\mathrm{O}(2d, \mathbb{R})$ stands of the group of $2d \times 2d$ orthogonal matrices. We also recall a result on a kind of singular value decomposition of symplectic matrices, also known as the *Euler decomposition* in the literature; see [206, Appendix B.2] for details and proofs.

Proposition 4.3.2 *For any $S \in \mathrm{Sp}(d, \mathbb{R})$ there exist $U, V \in \mathrm{U}(2d, \mathbb{R})$ such that*

$$S = U^\top D V, \quad D = \Sigma \oplus \Sigma^{-1},$$

where $\Sigma = \mathrm{diag}(\sigma_1, \ldots, \sigma_d)$ *and* $\sigma_1 \geq \ldots \geq \sigma_d \geq \sigma_d^{-1} \geq \ldots \geq \sigma_1^{-1}$ *are the singular values of S.*

It should be emphasized that the matrices U and V appearing in such factorization are generally not unique, owing to possibly degenerate singular values of S, whereas Σ is uniquely determined provided that a choice for the ordering of the singular values has been made. We can thus identify any Euler decomposition of S as $U^\top DV$ with the triple (U, V, Σ).

Let us introduce the notion of *free symplectic matrix.*

Definition 4.3.3 Let $S \in \mathrm{Sp}(d, \mathbb{R})$. We say that S is a free symplectic matrix if any of the following equivalent conditions is satisfied:

(i) If $S = \begin{bmatrix} A & B \\ C & D \end{bmatrix}$ then $\det B \neq 0$.

(ii) Given $(x, y) \in \mathbb{R}^{2d}$, there exists a unique $(\xi, \eta) \in \mathbb{R}^{2d}$ such that $(x, \xi) = S(y, \eta)$.

(iii) Set $(x, \xi) = S(y, \eta)$, $(y, \eta) \in \mathbb{R}^{2d}$. Then $\det(\partial_\eta x) \neq 0$.

The subset of free symplectic matrices is denoted by $\mathrm{Sp}_0(d, \mathbb{R})$.

Free symplectic matrices are naturally associated with quadratic forms. These are also called *generating functions*, in connection with those of canonical transformations in classical mechanics [73, 118].

Proposition 4.3.4 *Let* $S \in \mathrm{Sp}_0(d, \mathbb{R})$ *and define the generating function*

$$\Phi_S(x, y) := \frac{1}{2} DB^{-1}x \cdot x - B^{-1}x \cdot y + \frac{1}{2} B^{-1} Ay \cdot y. \qquad (4.4)$$

Therefore,

$$(x, \xi) = S(y, \eta) \iff \begin{cases} \xi = \nabla_x \Phi_S(x, y) \\ \eta = -\nabla_y \Phi_S(x, y). \end{cases}$$

Conversely, let $L, P, Q \in \mathbb{R}^{d \times d}$ *be such that* $P = P^\top$, $Q = Q^\top$ *and* $\det L \neq 0$, *and set*

$$\Phi(x, y) = \frac{1}{2} Px \cdot x - Lx \cdot y + \frac{1}{2} Qy \cdot y.$$

Then

$$S_\Phi := \begin{bmatrix} L^{-1}Q & L^{-1} \\ PL^{-1}Q - L^\top & PL^{-1} \end{bmatrix} \in \mathrm{Sp}_0(d, \mathbb{R}),$$

and its generating function in the sense above is Φ.

We finally recall a factorization result in terms of free matrices and provide a special set of generators of symplectic matrices.

Proposition 4.3.5 *(i) For any $S \in \mathrm{Sp}(d, \mathbb{R})$ there exist (non-unique) $S_1, S_2 \in$*
$\mathrm{Sp}_0(d, \mathbb{R})$ such that $S = S_1 S_2$.
(ii) Given $P, L \in \mathbb{R}^{d \times d}$ with $P = P^\top$ and $\det L \neq 0$, define

$$V_P := \begin{bmatrix} I & 0 \\ P & I \end{bmatrix}, \quad U_P := \begin{bmatrix} P & I \\ -I & 0 \end{bmatrix}, \quad M_L := \begin{bmatrix} L & 0 \\ 0 & (L^\top)^{-1} \end{bmatrix}. \tag{4.5}$$

If $S = \begin{bmatrix} A & B \\ C & D \end{bmatrix} \in \mathrm{Sp}_0(d, \mathbb{R})$ then

$$S = V_{DB^{-1}} M_B U_{B^{-1}A} = V_{DB^{-1}} M_B J V_{B^{-1}A}.$$

As a consequence, any of the sets $\{V_P, M_L, J : P = P^\top, \det L \neq 0\}$ and
$\{U_P, M_L : P = P^\top, \det L \neq 0\}$ generates $\mathrm{Sp}(d, \mathbb{R})$.

4.3.2 Metaplectic Operators: Definitions and Basic Properties

In this section we introduce a class of unitary operators in $L^2(\mathbb{R}^d)$ naturally associated with symplectic matrices. More precisely, with any symplectic matrix $S \in \mathrm{Sp}(d, \mathbb{R})$ we associate a set of unitary operators, called *metaplectic operators*, differing each other by a phase factor. We denote by $\mu(S)$ any one of such operators. They are usually defined as intertwining operators for the Schrödinger representation of the Heisenberg group, and the set of the operators $c\mu(S)$, $|c| = 1$, $S \in \mathrm{Sp}(d, \mathbb{R})$, constitutes the so-called metaplectic group $\mathrm{Mp}(d, \mathbb{R})$. Finally, it turns out that the map $S \mapsto \mu(S)$ is a projective unitary representation of $\mathrm{Sp}(d, \mathbb{R})$.

We are going to present a self-contained and direct construction of such operators, which avoids explicit reference to results from representation theory. In fact, a finer construction allows one to associate with any symplectic matrix S *two* operators, differing only by the sign, and one ends up with a representation of the double covering of $\mathrm{Sp}(d, \mathbb{R})$; comprehensive accounts on the topic can be found in [71, 106] but such a subtler analysis will not be relevant to the applications in this monograph.

Inspired by the spreading form of the Weyl quantization (4.2), we define the *Weyl time-frequency shifts* (also known as symmetric time-frequency shifts [121] or Weyl-Heisenberg operators [71]) by

$$\pi'(x, \xi) := M_{\xi/2} T_x M_{\xi/2} = e^{-\pi i x \cdot \xi} \pi(x, \xi), \quad (x, \xi) \in \mathbb{R}^{2d}.$$

Observe from (4.2) that

$$\pi'(z_0) = a_{z_0}^{\mathrm{w}}, \qquad a_{z_0}(z) = e^{-2\pi i J z \cdot z_0}, \qquad z, z_0 \in \mathbb{R}^{2d}. \tag{4.6}$$

It turns out that $\mu(S)$ satisfies the intertwining relation

$$\pi'(Sz) = \mu(S)\pi'(z)\mu(S)^{-1}, \quad z \in \mathbb{R}^{2d}, \tag{4.7}$$

which characterizes the unitary operator $\mu(S)$ up to a phase factor and consequently can be used to compute $\mu(S)$ (up to a phase factor).

In fact one can prove by explicit computation (cf. [121, Example 9.4.1] that the following choices do the job in the case of the special elements of $\mathrm{Sp}(d, \mathbb{R})$ highlighted in (4.5).

In the following we let $c \in \mathbb{C}$ be a phase factor, namely $|c| = 1$, which can be different from time to time.

1. The Fourier transform is a special metaplectic operator associated with the canonical symplectic matrix, that is $\mu(J)f = c\mathcal{F}(f)$. Notice in particular that $\mu(-J) = c\mathcal{F}^{-1}$.
2. Let $L \in \mathrm{GL}(d, \mathbb{R})$. The metaplectic operator $\mu(M_L)$ acts as a rescaling by L:

$$\mu(M_L)f(x) = c|\det L|^{-1/2}f(L^{-1}x).$$

3. Let $P \in \mathbb{R}^{d \times d}$ be a symmetric matrix. The metaplectic operator $\mu(V_P)$ is a chirp multiplication:

$$\mu(V_P)f(x) = ce^{\pi i x \cdot P x}f(x).$$

Actually, we can prove the following result.

Theorem 4.3.6 *For every $S \in \mathrm{Sp}(d, \mathbb{R})$ there exists a unitary operator $\mu(S)$ on $L^2(\mathbb{R}^d)$, unique up to a phase factor, satisfying (4.7).*

As a consequence, for every $S_1, S_2 \in \mathrm{Sp}(d, \mathbb{R})$ we have

$$\mu(S_1 S_2) = c\mu(S_1)\mu(S_2) \tag{4.8}$$

for some $c \in \mathbb{C}$, $|c| = 1$.

Proof The existence follows from the special cases discussed above, when S is equal to J, or of the form M_L or V_P, because such matrices generate $\mathrm{Sp}(d, \mathbb{R})$.

Concerning the uniqueness, we can argue as follows (cf. Schur's lemma in [159, Proposition 1.5]). Suppose that there exist two unitary operators A, B such that

$$\pi'(Sz) = A\pi'(z)A^{-1}, \quad \pi'(Sz) = B\pi'(z)B^{-1},$$

for some $S \in \mathrm{Sp}(d, \mathbb{R})$ and every $z \in \mathbb{R}^{2d}$. Then the operator $C := A^{-1}B$ commutes with $\pi'(z)$ (equivalently, with $\pi(z)$) for any $z \in \mathbb{R}^{2d}$. Since $\pi'(z)^* = c\pi(-z)$ for a suitable $c \in \mathbb{C}$, $|c| = 1$, the same conclusion holds for C^* and hence for

$$\mathrm{Re}\, C := \frac{C + C^*}{2}, \quad \mathrm{Im}\, C := \frac{C - C^*}{2i}.$$

Assume that A and B do not differ by a phase factor. As a result, $C \neq \lambda I$, $\lambda \in \mathbb{C}$ (if $|\lambda| \neq 1$ this is clear since C is unitary), and thus $\mathrm{Re}\, C \neq \lambda I$ or $\mathrm{Im}\, C \neq \lambda I$. Suppose without loss of generality that $\mathrm{Re}\, C \neq \lambda I$. The spectral theorem [190] applied to the bounded self-adjoint operator $\mathrm{Re}\, C$ implies the existence of a non-trivial spectral projection P (i.e., $P \neq 0$, $P \neq I$) which commutes with any shift $\pi(z)$, as $\mathrm{Re}\, C$ does. Let then $g \in \ker P \setminus \{0\}$ and $f \in (\ker P)^{\perp} \setminus \{0\}$; therefore $P\pi(z)g = \pi(z)Pg = 0$, so that $V_g f(z) = \langle f, \pi(z)g \rangle = 0$ for all $z \in \mathbb{R}^{2d}$. On the other hand, we have

$$\|V_g f\|_{L^2} = \|f\|_{L^2}\|g\|_{L^2} \neq 0,$$

which is a contradiction.

Finally, the projective representation property follows by the uniqueness up to a phase factor, since $\mu(S_1)\mu(S_2)$ is an intertwining operator as well as $\mu(S)$ if $S = S_1 S_2$.   □

Observe, as a consequence, that $\{\mu(V_P), \mu(M_L), J : P = P^{\top}, \det L \neq 0\}$ is a set of generators of $\mathrm{Mp}(d, \mathbb{R})$ (cf. [71, Corollary 112]).

Combining these results we obtain the following representation of metaplectic operators in terms of *quadratic Fourier transforms* (cf. [106, Theorems 4.51 and 4.53] and [71]).

Theorem 4.3.7 *Let* $S = \begin{bmatrix} A & B \\ C & D \end{bmatrix} \in \mathrm{Sp}(d, \mathbb{R})$.

(i) *If* $\det B \neq 0$ *then*

$$\mu(S)f(x) = c|\det B|^{-1/2} \int_{\mathbb{R}^d} e^{2\pi i \Phi_S(x,y)} f(y)dy, \qquad f \in S(\mathbb{R}^d), \qquad (4.9)$$

  for some $c \in \mathbb{C}$ *with* $|c| = 1$*, where* Φ_S *is the generating function of* S *defined in* (4.4).

(ii) *If* $\det A \neq 0$ *then*

$$\mu(S)f(x) = c|\det A|^{-1/2} \int_{\mathbb{R}^d} e^{2\pi i \Phi_{SJ^{-1}}(x,\eta)} \hat{f}(\eta)d\eta, \qquad f \in S(\mathbb{R}^d),$$

for some $c \in \mathbb{C}$ with $|c| = 1$, where

$$\Phi_{SJ^{-1}}(x, \eta) = \frac{1}{2}CA^{-1}x \cdot x + A^{-1}x \cdot \eta - \frac{1}{2}A^{-1}B\eta \cdot \eta.$$

Proof Concerning the first point, we use the factorization

$$S = V_{DB^{-1}} M_B J V_{B^{-1}A}$$

in Proposition 4.3.5. Then we apply the product formula (4.8) and the explicit expression for the metaplectic operators associated with matrices of the type V_P and M_L and J, as detailed before Theorem 4.3.6. The proof of the second point is similar.   □

4.3.3   The Schrödinger Equation with Quadratic Hamiltonian

A prominent example of a metaplectic operator is given by the Schrödinger propagator for the free particle $U(t) = e^{i(t/4\pi)\Delta}$, $t \in \mathbb{R}$. This characterization can be easily derived from the examples already discussed, since $U(t)$ is a Fourier multiplier with chirp symbol $m_t(\xi) = e^{-\pi i t |\xi|^2}$ on $\mathbb{R}^d$, hence

$$U(t) = \mathcal{F}^{-1} m_t \mathcal{F} = c(t)\mu(S_t), \quad S_t = \begin{bmatrix} I & tI \\ O & I \end{bmatrix} \in \mathrm{Sp}(d, \mathbb{R}), \quad t \in \mathbb{R},$$

where $c(t) \in \mathbb{C}$ satisfies $|c(t)| = 1$.

In general, let $Q: \mathbb{R}^{2d} \to \mathbb{R}$ be a homogeneous quadratic polynomial, namely

$$Q(x, \xi) = \frac{1}{2}A\xi \cdot \xi + Bx \cdot \xi + \frac{1}{2}Cx \cdot x,$$

for some $A, B, C \in \mathbb{R}^{d \times d}$ with $A = A^\top$ and $C = C^\top$. Let us consider the Schrödinger equation (with the normalization $\hbar = 1/2\pi$)

$$\begin{cases} \frac{i}{2\pi} \partial_t \psi = Q^w \psi \\ \psi(0, x) = f(x), \end{cases} \tag{4.10}$$

where the Hamiltonian is the Weyl quantization of Q, namely, with $D = (D_1, \ldots, D_d)$, $D_j = (2\pi i)^{-1}\partial_j$,

$$
\begin{aligned}
Q^{\mathrm{w}} &= \frac{1}{2}AD \cdot D + \frac{1}{2}(Bx \cdot D + D \cdot Bx) + \frac{1}{2}Cx \cdot x \\
&= \frac{1}{2}AD \cdot D + Bx \cdot D + \frac{1}{2}Cx \cdot x - \frac{i}{4\pi}\mathrm{Tr}(B) \\
&= -\frac{1}{8\pi^2}\sum_{j,k=1}^{d} A_{j,k}\partial_j\partial_k - \frac{i}{2\pi}\sum_{j,k=1}^{d} B_{j,k}x_j\partial_k + \frac{1}{2}\sum_{j,k=1}^{d} C_{j,k}x_j x_k - \frac{i}{4\pi}\mathrm{Tr}(B).
\end{aligned}
$$

Examples of systems encompassed by such Hamiltonians are: a non-relativistic spinless particle in a time-independent and uniform magnetic field; the standard harmonic oscillator and its anisotropic variants.

Theorem 4.3.8 *Using the notation introduced so far:*

(i) *For every $f \in S(\mathbb{R}^d)$ the Cauchy problem (4.10) has a unique solution $\psi \in C^\infty(\mathbb{R}; S(\mathbb{R}^d))$. Moreover, the propagator $U(t)$ defined by $U(t)f = \psi(t, \cdot)$ is continuous on $S(\mathbb{R}^d)$ for any $t \in \mathbb{R}$ and satisfies the group property $U(t)U(s) = U(t + s)$, $t, s \in \mathbb{R}$.*

(ii) *The operator $U(t)$ extends continuously to a one-parameter strongly continuous group of unitary operators on $L^2(\mathbb{R}^d)$. If $f \in L^2(\mathbb{R}^d)$ then $U(t)f$ belongs to $C_b(\mathbb{R}; L^2(\mathbb{R}^d))$ and satisfies the equation in the sense of temperate distributions in $\mathbb{R} \times \mathbb{R}^d$.*

(iii) *The operator $U(t)$ extends to a one-parameter strongly smooth group of automorphisms of $S'(\mathbb{R}^d)$. Therefore, if $f \in S'(\mathbb{R}^d)$ then $U(t)f$ belongs to $C^\infty(\mathbb{R}; S'(\mathbb{R}^d))$ and satisfies the equation in the distribution sense. Moreover, it is the only solution in $C(\mathbb{R}; S'(\mathbb{R}^d))$ of the problem (4.10).*

Proof (i) We look for a solution in the form

$$
\psi(t, x) = U(t)f(x) = g(t)\int_{\mathbb{R}^d} e^{2\pi i \Phi_t(x,\eta)} \hat{f}(\eta)d\eta,
$$

for some real quadratic form Φ_t with coefficients depending smoothly on t and some smooth factor $g(t)$. A formal substitution in the equation yields

$$
\frac{1}{2\pi i}\partial_t\psi(t, x) + (Q^{\mathrm{w}}\psi)(t, x)
$$
$$
= \int_{\mathbb{R}^d} e^{2\pi i \Phi_t(x,\eta)}\left(\partial_t\Phi_t(x, \eta) + Q(x, \nabla_x\Phi_t(x, \eta)) + \frac{1}{2\pi i}g'(t) + \alpha(t)\right)\hat{f}(\eta)d\eta,
$$

where

$$\alpha(t) = -\frac{i}{4\pi}\mathrm{Tr}(B) + \frac{1}{2\pi i}\sum_{j,k=1}^{d} A_{j,k}\partial_j\partial_k\Phi_t(x,\eta)$$

is indeed a function of t only. We now require that Φ_t satisfies the Hamilton-Jacobi equation

$$\partial_t\Phi_t(x,\eta) + Q(x, \nabla_x\Phi_t(x,\eta)) = 0, \qquad (4.11)$$

as well as $\Phi_0(x,\eta) = x \cdot \eta$, and then that

$$\frac{1}{2\pi i}g'(t) + \alpha(t) = 0, \qquad g(0) = 1.$$

Since both Q and Φ_t are quadratic forms, it follows that (4.11) provides an autonomous nonlinear system of ordinary differential equations for the coefficients of Φ_t. The Cauchy theorem implies that there exists a unique smooth solution, at least for $t \in (-\epsilon, \epsilon)$ for some $\epsilon > 0$. We have then constructed $U(t)$, $t \in (-\epsilon, \epsilon)$, in the desired form. By expanding $e^{2\pi i\Phi_t}$ as a product, we see that $U(t)$ can be written (except for a smooth factor) as the composition of operators of the type $\mu(J)$, $\mu(M_L)$ and $\mu(V_P)$ introduced before Theorem 4.3.6, for suitable matrices P, L whose entries depend smoothly on $t \in (-\epsilon, \epsilon)$. As a consequence, we see that $U(t)$ is continuous on $\mathcal{S}(\mathbb{R}^d)$ for every $t \in (-\epsilon, \epsilon)$ and the map $t \mapsto U(t)$ is strongly smooth, that is $\psi(t, \cdot) = U(t)f \in C^\infty((-\epsilon, \epsilon), \mathcal{S}(\mathbb{R}^d))$ if $f \in \mathcal{S}(\mathbb{R}^d)$. Finally, note that $\psi = U(t)f$, $f \in \mathcal{S}(\mathbb{R}^d)$, solves the equation in (4.10) in $(-\epsilon, \epsilon) \times \mathbb{R}^d$ in the classical sense, since the formal manipulation above can be justified by the dominated convergence theorem.

Since the equation in (4.10) is autonomous, if $f \in \mathcal{S}(\mathbb{R}^d)$ then $U(t-s)f$ is a solution for $t \in (s-\epsilon, s+\epsilon)$. We can thus construct a global solution $\psi \in C^\infty(\mathbb{R}; \mathcal{S}(\mathbb{R}^d))$ of the Cauchy problem (4.10) by gluing together local solutions, defined on consecutive closed intervals of suitable length, say $\epsilon/2$. The smoothness of ψ at the common endpoints of any couple of adjacent intervals can be proved by induction on the order of the time derivatives, using the equation recursively.

We now set $U(t)f = \psi(t, \cdot)$, $t \in \mathbb{R}$, for $f \in \mathcal{S}(\mathbb{R}^d)$. To prove that $U(t)f$ is the unique solution of (4.10) we observe that if $\psi \in C^1(\mathbb{R}; \mathcal{S}(\mathbb{R}^d))$ is a solution then

$$\begin{aligned}
\frac{d}{dt}\|\psi(t, \cdot)\|_{L^2}^2 &= 2\mathrm{Re}\langle\psi(t, \cdot), \partial_t\psi(t, \cdot)\rangle \\
&= -4\pi\mathrm{Re}(i\langle\psi(t, \cdot), Q^w\psi(t, \cdot)\rangle) \\
&= 0,
\end{aligned}$$

because Q is a real-valued symbol, so that Q^w is symmetric on $\mathcal{S}(\mathbb{R}^d)$. Hence we obtain the energy estimate

$$\|\psi(t,\cdot)\|_{L^2} = \|f\|_{L^2} \quad \forall t \in \mathbb{R} \quad (f = \psi(0,\cdot)), \tag{4.12}$$

that implies uniqueness in $C^1(\mathbb{R}; \mathcal{S}(\mathbb{R}^d))$ at once.

The group property of $U(t)$ follows from uniqueness, because $U(t+s)f$ and $U(t)U(s)f$ for fixed $f \in \mathcal{S}(\mathbb{R}^d)$ and $s \in \mathbb{R}$ are solutions in $C^\infty(\mathbb{R}; \mathcal{S}(\mathbb{R}^d))$ of the same problem.

(ii) The energy estimate (4.12) implies that $U(t)$ extends to an isometry on $L^2(\mathbb{R}^d)$ and, by continuity, the group property $U(t)U(s) = U(t+s)$, $s, t \in \mathbb{R}$, still holds in L^2. In particular, $U(t)$ is invertible, hence a unitary operator. Finally, if $f \in L^2(\mathbb{R}^d)$, let $f_n \in \mathcal{S}(\mathbb{R}^d)$ be such that $f_n \to f$ in $L^2(\mathbb{R}^d)$ and let $\psi_n = U(t)f_n$ be the corresponding smooth solutions, hence satisfying (4.12). Then ψ_n is a Cauchy sequence in $C_b(\mathbb{R}; L^2(\mathbb{R}^d))$ and its limit must coincide with $U(t)f$, because $U(t)f_n \to U(t)f$ for all $t \in \mathbb{R}$. Therefore, $U(t)f$ is the limit in the sense of (temperate) distributions of solutions; hence it is a distribution solution.

(iii) For $f \in \mathcal{S}'(\mathbb{R}^d)$, $t \in \mathbb{R}$, we define $U(t)f \in \mathcal{S}'(\mathbb{R}^d)$ by

$$\langle U(t)f, \phi \rangle = \langle f, U(-t)\phi \rangle, \quad \phi \in \mathcal{S}(\mathbb{R}^d).$$

Then $U(t)$ is a continuous extension of the propagator defined on $\mathcal{S}(\mathbb{R}^d)$. The group property clearly extends to $\mathcal{S}'(\mathbb{R}^d)$. Moreover, if $f \in \mathcal{S}'(\mathbb{R}^d)$ then $U(t)f \in C^\infty(\mathbb{R}; \mathcal{S}'(\mathbb{R}^d))$. Now, if $f_n \to f$ in $\mathcal{S}'(\mathbb{R}^d)$ then $U(t)f_n \to U(t)f$ in $C(\mathbb{R}; \mathcal{S}'(\mathbb{R}^d))$; indeed, for any $T > 0$ and $\phi \in \mathcal{S}(\mathbb{R}^d)$, we have

$$\sup_{t \in [-T,T]} |\langle f_n - f, U(-t)\phi \rangle| \to 0,$$

since $f_n \to f$ in $\mathcal{S}'(\mathbb{R}^d)$ for the topologies of pointwise and bounded convergence—that is the same, cf. [223, Proposition 34.4 and Corollary 1, page 358]—hence uniformly on the compact subset $\{U(-t)\phi : t \in [-T, T]\} \subset \mathcal{S}(\mathbb{R}^d)$. This shows that $U(t)f$, $f \in \mathcal{S}'(\mathbb{R}^d)$, is a well-defined distribution in $\mathbb{R} \times \mathbb{R}^d$ and is the limit in $C(\mathbb{R}; \mathcal{S}'(\mathbb{R}^d)) \hookrightarrow \mathcal{D}'(\mathbb{R} \times \mathbb{R}^d)$ of smooth solutions, hence it is a solution as well.

Finally, let $\psi \in C^0(\mathbb{R}; \mathcal{S}'(\mathbb{R}^d))$ be a distribution solution of the problem (4.10). Since Q^w is continuous on $\mathcal{S}'(\mathbb{R}^d)$, it follows from the equation that $\psi \in C^1(\mathbb{R}; \mathcal{S}'(\mathbb{R}^d))$ and therefore $U(-t)\psi(t,\cdot) \in C^1(\mathbb{R}; \mathcal{S}'(\mathbb{R}^d))$ and

$$\frac{d}{dt}U(-t)\psi(t,\cdot) = 0,$$

which implies that $\psi(t,\cdot) = U(t)\psi(0) = U(t)f$. Here we implicitly used the fact that Q^w commutes with $U(t)$ on $\mathcal{S}'(\mathbb{R}^d)$. In fact, it is sufficient to check the latter on $\mathcal{S}(\mathbb{R}^d)$, which in turn follows from the fact that both $Q^w U(t) f$ and $U(t) Q^w f$, $f \in \mathcal{S}(\mathbb{R}^d)$, are solutions in $C^\infty(\mathbb{R}; \mathcal{S}(\mathbb{R}^d))$ of the same Cauchy problem.

$\square$

We will use the notation $U(t) = e^{-2\pi i t Q^w}$ for the propagator just constructed. It follows from the proof of the above result that $U(t)$ is in fact a metaplectic operator. We now prove that it is associated to the symplectic matrix defined by the corresponding Hamiltonian flow. In fact, the classical phase-space flow determined by the Hamilton equations

$$\dot{z} = J\nabla_z Q(z) = \begin{bmatrix} B & A \\ -C & -B^\top \end{bmatrix} \begin{bmatrix} x \\ y \end{bmatrix} =: \mathbb{S}z, \quad z = (x, y) \in \mathbb{R}^{2d},$$

is given by the mapping $\mathbb{R} \ni t \mapsto S_t := e^{t\mathbb{S}} \in \mathrm{Sp}(d, \mathbb{R})$. The following result clarifies the connection between S_t and the propagator $U(t) = e^{-2\pi i t Q^w}$ (cf. [71, Section 15.1.3]).

Theorem 4.3.9 *With the notation introduced above,*

$$\mu(S_t) = c(t) e^{-2\pi i t Q^w}, \tag{4.13}$$

for some $c(t) \in \mathbb{C}$, $|c(t)| = 1$.

Proof Using the uniqueness result in Theorem 4.3.6 and the fact that the Weyl time-frequency shifts $\pi'(z)$ are Weyl operators (cf. (4.6)), it is sufficient to prove that

$$e^{2\pi i t Q^w} a^w e^{-2\pi i t Q^w} = (a \circ S_t)^w, \tag{4.14}$$

for any $t \in \mathbb{R}$ and $a \in \mathcal{S}'(\mathbb{R}^{2d})$. Indeed, one can then take $a(z) = a_{z_0}(z) = e^{-2\pi i J z \cdot z_0}$, $z, z_0 \in \mathbb{R}^{2d}$, and use that $a_{z_0}(S_t z) = a_{S_t^{-1} z_0}(z)$, since $J S_t = (S_t^\top)^{-1} J$. We know from Theorem 4.3.8 that the propagator $e^{-2\pi i t Q^w}$ defines a C^∞ one-parameter group of automorphisms of $\mathcal{S}(\mathbb{R}^d)$ and $\mathcal{S}'(\mathbb{R}^d)$. Therefore, the left-hand side of (4.14) defines a linear bounded operator $\mathcal{S}(\mathbb{R}^d) \to \mathcal{S}'(\mathbb{R}^d)$, so that it has a Weyl symbol that we denote by $F_t(a)$. Similarly, we define $G_t(a) = a \circ S_t$ and we are required to prove that $F_t(a) = G_t(a)$ for any $a \in \mathcal{S}'(\mathbb{R}^{2d})$ and $t \in \mathbb{R}$.

Observe that the maps $t \mapsto F_t$ and $t \mapsto G_t$ are continuous in $\mathcal{L}_s(\mathcal{S}'(\mathbb{R}^{2d}))$, namely for any $a \in \mathcal{S}'(\mathbb{R}^{2d})$ the map $t \mapsto F_t(a)$ is continuous in $\mathcal{S}'(\mathbb{R}^{2d})$—the same holds for $t \mapsto G_t(a)$. In fact this is equivalent to the continuity of the map $t \mapsto \langle F_t(a)^w f, g \rangle$ for every $f, g \in \mathcal{S}(\mathbb{R}^d)$, and similarly for G_t, by Proposition 6.2.1 below, because the map which associates a Weyl symbol with the corresponding distribution kernel is an isomorphism of $\mathcal{S}'(\mathbb{R}^{2d})$. Moreover, we have $F_{t_1} \circ F_{t_2} = F_{t_1+t_2}$ for every $t_1, t_2 \in \mathbb{R}$ and similarly for G_t.

We will prove in a moment that

$$F'_t|_{t=0} = G'_t|_{t=0} =: A, \tag{4.15}$$

where again the derivatives are understood in $\mathcal{L}_s(S'(\mathbb{R}^{2d}))$, i.e., for every $a \in S'(\mathbb{R}^d)$,

$$\frac{1}{h}(F_h(a) - F_0(a)) \to A(a)$$

in $S'(\mathbb{R}^d)$, as $t \to 0$, or equivalently

$$\frac{d}{dt}\langle (F_t(a))^w f, g\rangle|_{t=0} = \langle (A(a))^w f, g\rangle$$

for every $a \in S'(\mathbb{R}^d)$ $f, g \in S(\mathbb{R}^d)$, and similarly for G_t. Note that A commutes with F_t for every $t \in \mathbb{R}$, because it is the limit of maps commuting with F_t—the same clearly holds for G_t too. As a consequence of the group properties mentioned above, we see that F_t and G_t are differentiable for all $t \in \mathbb{R}$ and satisfy the equations $F'_t = A \circ F_t$, $G'_t = A \circ G_t$ in $\mathcal{L}_s(S'(\mathbb{R}^{2d}))$, so that

$$(F_{-t} \circ G_t)' = -A \circ F_{-t} \circ G_t + F_t \circ A \circ G_t = 0.$$

Since $F_0 = G_0 = I$ we infer $F_{-t} \circ G_t = I$, hence $F_t = F_t \circ F_{-t} \circ G_t = G_t$ as desired. The above application of the Leibniz rule in $\mathcal{L}_s(S'(\mathbb{R}^d))$ is justified as usual by writing the increment as

$$F_{-(t+h)} \circ G_{t+h} - F_{-t} \circ G_t = F_{-(t+h)} \circ (G_{t+h} - G_t) + (F_{-(t+h)} - F_t) \circ G_t,$$

and using the fact that the composition of maps $\mathcal{L}_s(S'(\mathbb{R}^d)) \times \mathcal{L}_s(S'(\mathbb{R}^d)) \to \mathcal{L}_s(S'(\mathbb{R}^d))$ is bilinear sequentially continuous—this latter fact can be easily verified using the Banach-Steinhaus theorem [223, Theorem 33.1], which holds for maps on $S'(\mathbb{R}^d)$, being $S'(\mathbb{R}^d)$ barreled [223, Page 376].

Hence, it remains to prove (4.15). On the one hand, we have

$$(F'_t|_{t=0}(a))^w = 2\pi i[Q^w, a^w],$$

and since Q is a second degree polynomial we infer from the Weyl calculus of pseudodifferential operators [106, Corollary 2.51] that

$$[Q^w, a^w] = \frac{1}{2\pi i}(J\nabla Q \cdot \nabla a)^w$$

(notice that $J\nabla Q \cdot \nabla a$ is the Poisson bracket of Q and a), hence

$$F'_t|_{t=0}(a) = J\nabla Q \cdot \nabla a.$$

On the other hand, by the chain rule we have

$$G'_t|_{t=0}(a) = \frac{d}{dt}(S_t z) \cdot \nabla a = J \nabla Q \cdot \nabla a,$$

and the proof of (4.15) is concluded.   □

Example 4.3.10 Consider the classical Hamiltonian $Q(x, \xi) = \frac{1}{2}|\xi|^2 + \frac{1}{2}|x|^2$ and the corresponding Schrödinger equation.

$$\frac{i}{2\pi}\partial_t \psi = -\frac{1}{8\pi^2}\Delta \psi + \frac{1}{2}|x|^2 \psi.$$

The classical flow S_t is given by

$$S_t = \begin{bmatrix} (\cos t)I & (\sin t)I \\ -(\sin t)I & (\cos t)I \end{bmatrix}.$$

Using Theorems 4.3.9 and 4.3.7 we obtain the distribution kernel $u_t(x, y)$ of the corresponding evolution operator $U(t)$, known as the *Mehler kernel*: for $k \in \mathbb{Z}$,

$$u_t(x, y) = \begin{cases} c(k)|\sin t|^{-d/2} \exp\left(\pi i \frac{x^2+y^2}{\tan t} - 2\pi i \frac{x \cdot y}{\sin t}\right) & (\pi k < t < \pi(k+1)) \\ c'(k)\delta((-1)^k x - y) & (t = k\pi), \end{cases}$$

for suitable phase factors $c(k), c'(k) \in \mathbb{C}$; cf. [71, 155]. Observe that for $t \neq k\pi, k \in \mathbb{Z}$, u_t is a smooth function. We remark that the standard physics textbook derivation of the integral formula in (4.9) involves Feynman path integrals. In this respect, it is worth to emphasize that (4.9) can be actually interpreted as a rigorous definition of path integral in the case of a quadratic Hamiltonian, cf. [73, 193].

4.3.4 Symplectic Covariance of the Weyl Calculus

The Weyl quantization satisfies a special intertwining property involving metaplectic operators, called *symplectic covariance* (cf. [71, Theorem 215]).

Proposition 4.3.11 *Let $\mu(S)$ be a metaplectic operator associated with $S \in Sp(d, \mathbb{R})$.*

(i) For every $\sigma \in S'(\mathbb{R}^{2d})$ we have

$$(\sigma \circ S)^{\mathrm{w}} = \mu(S)^{-1} \sigma^{\mathrm{w}} \mu(S). \tag{4.16}$$

(ii) For every $f, g \in S(\mathbb{R}^d)$ we have

$$W(\mu(S)f, \mu(S)g) = W(f, g) \circ S^{-1}. \tag{4.17}$$

Proof

(i) Using the spreading representation (4.2) we express the Weyl quantization as a superposition of Weyl time-frequency shifts, then the covariance property (4.7) of the latter yields

$$\mu(S)^{-1}\sigma^{w}\mu(S) = \int_{\mathbb{R}^{2d}} \widehat{\sigma}(Jz)\mu(S)^{-1}\pi'(z)\mu(S)dz$$

$$= \int_{\mathbb{R}^{2d}} \widehat{\sigma}(Jz)\pi'(S^{-1}z)dz$$

$$= \int_{\mathbb{R}^{2d}} \widehat{\sigma}(JSz)\pi'(z)dz$$

$$= \int_{\mathbb{R}^{2d}} \widehat{\sigma}((S^{\top})^{-1}Jz)\pi'(z)dz$$

$$= \int_{\mathbb{R}^{2d}} \widehat{\sigma \circ S}(Jz)\pi'(z)dz$$

$$= (\sigma \circ S)^{w},$$

where we used that $\det S = 1$, $JS = (S^{\top})^{-1}J$ (recall that $S \in \mathrm{Sp}(d, \mathbb{R})$) and $\widehat{\sigma \circ S} = \widehat{\sigma} \circ (S^{\top})^{-1}$.

(ii) Formula (4.16) implies that, for every $\sigma \in S'(\mathbb{R}^{2d})$,

$$\langle \sigma, W(\mu(S)f, \mu(S)g) \rangle = \langle \sigma^{w}\mu(S)g, \mu(S)f \rangle$$

$$= \langle (\sigma \circ S)^{w}g, f \rangle$$

$$= \langle \sigma \circ S, W(f, g) \rangle$$

$$= \langle \sigma, W(f, g) \circ S^{-1} \rangle.$$

□

Symplectic covariance is in fact a distinctive property of the Weyl quantization among all possible quantization rules, as detailed below [63, 240].

Theorem 4.3.12 *Let* op$: S'(\mathbb{R}^{2d}) \to \mathcal{L}(S(\mathbb{R}^d), S'(\mathbb{R}^d))$ *be a continuous linear operator such that:*

1. *if $\sigma(x, \xi) = m(x)$ and $m \in L^{\infty}(\mathbb{R}^d)$ then* op$(\sigma)f(x) = m(x)f(x)$;
2. *for any $S \in \mathrm{Sp}(d, \mathbb{R})$ and $\sigma \in S'(\mathbb{R}^{2d})$,* op$(\sigma \circ S) = \mu(S)^{-1}op(\sigma)\mu(S)$.

Then, op$(\sigma) = \sigma^{w}$ *for every $\sigma \in S'(\mathbb{R}^d)$.*

4.3.5 Gabor Matrix of Metaplectic Operators

Metaplectic operators have been extensively studied through the lens of Gabor analysis. The purpose of this section is to collect some of the most important results concerning the Gabor matrix of a metaplectic operator and the boundedness on modulation spaces—see [54] for further details.

Theorem 4.3.13 *Consider* $\mu(S) \in \mathrm{Mp}(d, \mathbb{R})$ *and* $g, \gamma \in \mathcal{S}(\mathbb{R}^d)$. *For any* $N \geq 0$ *we have*

$$|\langle \mu(S)\pi(z)g, \pi(w)\gamma \rangle| \lesssim_{N,S} (1 + |w - Sz|)^{-N}, \quad w, z \in \mathbb{R}^{2d}.$$

As a consequence, for any $1 \leq p \leq \infty$ *and* $s \in \mathbb{R}$, *the operator* $\mu(S)$ *is bounded from* $M_{v_s}^p(\mathbb{R}^d)$ *into itself.*

The continuity of $\mu(S)$ on $M^{p,q}(\mathbb{R}^d)$ with $p \neq q$ fails in general, cf. [53, Proposition 7.1].

Proof We use the Moyal formula (3.5), the covariance properties of Wigner distribution (3.4) and (4.17). Hence

$$|\langle \mu(S)\pi(z)g, \pi(w)\gamma \rangle|^2 = \int_{\mathbb{R}^{2d}} W(\mu(S)\pi(z)g)(u)W(\pi(w)\gamma)(u)du$$

$$= \int_{\mathbb{R}^{2d}} W(\pi(z)g)(S^{-1}u)W\gamma(u - w)du$$

$$= \int_{\mathbb{R}^{2d}} Wg(S^{-1}u - z)W\gamma(u - w)du$$

$$= \int_{\mathbb{R}^{2d}} Wg(S^{-1}u + S^{-1}w - z)W\gamma(u)du.$$

Direct application of Proposition 3.1.3 (ii) yields, for any $s \geq 0$,

$$|\langle \mu(S)\pi(z)g, \pi(w)\gamma \rangle|^2 \lesssim \int_{\mathbb{R}^{2d}} v_{-s}(w - Sz - u)v_{-s}(u)du.$$

The claim follows since $v_{-s} * v_{-s} \lesssim_s v_{-s}$, cf. [121, Lemma 11.1.1(c)], for any $s > 2d$. $\square$

Remark 4.3.14 It is worth noting that, in light of Theorem 4.3.13, the Gabor matrix $\langle \mu(S)\pi(z)g, \pi(w)\gamma \rangle$ of a metaplectic operator $\mu(S) \in \mathrm{Mp}(d, \mathbb{R})$ is well defined in the case where

$$g \in M^p(\mathbb{R}^d), \ \gamma \in M^q(\mathbb{R}^d), \quad \frac{1}{p} + \frac{1}{q} \geq 1. \tag{4.18}$$

To be precise,

$$\|\mu(S)\pi(z)g\|_{M^p} \leq \|\mu(S)\|_{M^p \to M^p}\|\pi(z)g\|_{M^p} = \|\mu(S)\|_{M^p \to M^p}\|g\|_{M^p}, \quad z \in \mathbb{R}^d,$$

hence by Proposition 3.2.3 (iv)

$$
\begin{aligned}
|\langle \mu(S)\pi(z)g, \pi(w)\gamma \rangle| &\leq \|\mu(S)\|_{M^p \to M^p}\|g\|_{M^p}\|\pi(w)\gamma\|_{M^{p'}} \\
&\leq \|\mu(S)\|_{M^p \to M^p}\|g\|_{M^p}\|\gamma\|_{M^{p'}} \\
&\leq \|\mu(S)\|_{M^p \to M^p}\|g\|_{M^p}\|\gamma\|_{M^q},
\end{aligned}
$$

since from (4.18) we infer $q \leq p'$ and the inclusion $M^q(\mathbb{R}^d) \subset M^{p'}(\mathbb{R}^d)$ (Proposition 3.2.3 (v)) yields the last inequality.

4.4 Fourier and Oscillatory Integral Operators

4.4.1 Canonical Transformations and the Associated Operators

We now introduce a parametric family of algebras consisting of Fourier integral operators, characterized in terms of sparsity with respect to Gabor wave packets. This is a brief outline of the results developed in the articles [54, 55, 214].

First, we say that a canonical transformation $(x, \xi) = \chi(y, \eta)$ is *tame* (cf. (1.13) and (1.14)) if it does enjoy the following properties:

A1. $\chi : \mathbb{R}^{2d} \to \mathbb{R}^{2d}$ is smooth, invertible, and preserves the standard symplectic form in $\mathbb{R}^{2d}$, i.e., $dx \wedge d\xi = dy \wedge d\eta$; χ is a *symplectomorphism*.

A2. We have

$$|\partial_y^\alpha \partial_\eta^\beta \chi(y, \eta)| \leq C_{\alpha,\beta}, \quad |\alpha| + |\beta| \geq 1, \ y, \eta \in \mathbb{R}^d. \tag{4.19}$$

For example, the canonical transformation $\chi(t, s)$ in (1.13) satisfies *A1* and *A2* with constants $C_{\alpha,\beta} = C_{\alpha,\beta}(T_0)$, provided $|t - s| \leq T_0$ ($T_0 > 0$ being arbitrary), in view of (1.14).

We remark that *A1* and *A2* imply that χ and χ^{-1} are globally Lipschitz, hence

$$C^{-1}\langle w - \chi(z) \rangle \leq \langle \chi^{-1}(w) - z \rangle \leq C\langle w - \chi(z) \rangle, \quad w, z \in \mathbb{R}^{2d},$$

for some constant $C > 0$ *depending only on an upper bound for the first derivatives of* χ. This property will be used below with the canonical transformation $\chi = \chi(t, s)$ in (1.13), therefore with a constant $C = C(T_0)$ for $|t - s| \leq T_0$.

The following class of operators was introduced in [214].

Definition 4.4.1 Let χ be a tame canonical transformation. Let $g \in \mathcal{S}(\mathbb{R}^d) \setminus \{0\}$ and $s \geq 0$. We denote by $FIO(\chi, v_s)$ the class of operators $T: \mathcal{S}(\mathbb{R}^d) \to \mathcal{S}'(\mathbb{R}^d)$ such that

$$|\langle T\pi(z)g, \pi(w)g \rangle| \leq C_s v_{-s}(w - \chi(z)), \quad z, w \in \mathbb{R}^{2d}.$$

We also define

$$FIO(\chi) := \bigcap_{m \geq 0} FIO(\chi, v_m).$$

The latter space is endowed with the family of seminorms

$$\|T\|_{m,\chi} = \sup_{z,w \in \mathbb{R}^{2d}} v_m(w - \chi(z))|\langle T\pi(z)g, \pi(w)g \rangle|, \quad m \geq 0.$$

It is proved in [54] that the definition of $FIO(\chi)$ does not depend on the window $g \in \mathcal{S}(\mathbb{R}^d) \setminus \{0\}$.

The following two theorems can also be found in [54], but they were proved there in a discrete framework. We sketch below a proof following [179], where we highlight further properties—for instance, we focus on the uniformity of the constants.

Theorem 4.4.2 *Let $T \in FIO(\chi)$. Then T extends to a bounded operator on $M^p(\mathbb{R}^d)$, $1 \leq p \leq \infty$ (and in particular on $L^2(\mathbb{R}^d) = M^2(\mathbb{R}^d)$). Moreover, for every $m > 2d$ there exists a constant $C > 0$ depending only on m and the dimension d such that*

$$\|T\|_{M^p \to M^p} \leq C\|T\|_{m,\chi}.$$

Proof In view of the general strategy outlined in Sect. 4.1 (with $g = \gamma \in \mathcal{S}(\mathbb{R}^d) \setminus \{0\}$ for simplicity), since $V_g: M^p(\mathbb{R}^d) \to L^p(\mathbb{R}^{2d})$ and $V_g^*: L^p(\mathbb{R}^{2d}) \to M^p(\mathbb{R}^d)$ are bounded operators it is enough to prove that the operator $\widetilde{T} = V_g T V_g^*$ is bounded on $L^p(\mathbb{R}^{2d})$. Recall from (4.1) that this in an integral operator in $\mathbb{R}^{2d}$ with kernel

$$K_T(w, z) = \langle T\pi(z)g, \pi(w)g \rangle.$$

Since $T \in FIO(\chi)$ we have, for every $m \in \mathbb{N}$,

$$|\widetilde{T} F(w)| \leq \|T\|_{m,\chi} \int_{\mathbb{R}^{2d}} v_{-m}(w - \chi(z))|F(z)| \, dz.$$

Taking $m > 2d$, the claimed continuity on $L^p(\mathbb{R}^{2d})$ is proved by combining Schur's test and the fact that the Jacobian determinant of χ is $= 1$, χ being symplectic. $\quad\square$

Theorem 4.4.3 *If $T^{(i)} \in FIO(\chi_i)$, $i = 1, 2$, then the composition $T^{(1)}T^{(2)}$ is in $FIO(\chi_1 \circ \chi_2)$. Furthermore, for every $m > 2d$ there exists a constant $C > 0$ which depends only on m, the dimension d, and upper bounds for the first derivatives of χ_1, such that*

$$\|T^{(1)}T^{(2)}\|_{m,\chi_1 \circ \chi_2} \leq C\|T^{(1)}\|_{m,\chi_1}\|T^{(2)}\|_{m,\chi_2}.$$

Proof By resorting once more to the results in Sect. 4.1, we have to estimate the integral kernel of $V_g T^{(1)}T^{(2)}V_g^*$, where $g \in S(\mathbb{R}^d)$ satisfies $\|g\|_{L^2} = 1$. We write

$$V_g T^{(1)}T^{(2)}V_g^* = V_g T^{(1)}V_g^* V_g T^{(2)}V_g^*,$$

so that for $m > 2d$

$$|\langle T^{(1)}T^{(2)}\pi(z)g, \pi(w)g\rangle|$$

$$\leq \|T^{(1)}\|_{m,\chi_1}\|T^{(2)}\|_{m,\chi_2}\int_{\mathbb{R}^{2d}} v_{-m}(w - \chi_1(\zeta)) \times v_{-m}(\zeta - \chi_2(z))\, d\zeta$$

$$\leq C\|T^{(1)}\|_{m,\chi_1}\|T^{(2)}\|_{m,\chi_2}\int_{\mathbb{R}^{2d}} v_{-m}(w - \chi_1(\zeta))v_{-m}(\chi_1(\zeta) - \chi_1 \circ \chi_2(z))d\zeta.$$

The change of variable $\tilde{\zeta} = \chi_1(\zeta) - \chi_1 \circ \chi_2(z)$ and the convolution inequality $v_{-m} * v_{-m} \lesssim v_{-m}$ for $m > 2d$ (see [121, Lemma 11.1.1(c)]) give the desired estimate. $\square$

The reader can appreciate the simplicity of this result as compared to the composition formula obtained in [109, Theorem A.2] for oscillatory integral operators, whose proof really represents a technical *tour de force*—moreover, the result holds only in a short time setting.

For the sake of both illustration and future reference, we introduce here a family of Fourier integral operators belonging to $FIO(\chi)$ for a suitable χ. We say that a real phase function Φ on $\mathbb{R}^{2d}$ is *tame* if the following properties are satisfied:

B1. $\Phi \in C^\infty(\mathbb{R}^{2d})$;
B2. We have

$$|\partial_x^\alpha \partial_\eta^\beta \Phi(x, \eta)| \leq C_{\alpha,\beta}, \quad |\alpha| + |\beta| \geq 2, \ x, \eta \in \mathbb{R}^d; \quad (4.20)$$

B3. There exists $\tilde{\delta} > 0$ such that

$$|\det \partial_{x,\eta}^2 \Phi(x, \eta)| \geq \tilde{\delta}, \quad x, \eta \in \mathbb{R}^d. \quad (4.21)$$

Setting

$$\begin{cases} y = \nabla_\eta \Phi(x, \eta) \\ \xi = \nabla_x \Phi(x, \eta), \end{cases}$$

we can solve with respect to (x, ξ) resorting to Hadamard's global inverse function theorem [41, Theorem 2, page 93]. We thus obtain a mapping χ defined by $(x, \xi) = \chi(y, \eta)$ and satisfying the conditions $A1$, $A2$ above, as well as the following property:

$A3$. There exists $\tilde{\delta} > 0$ such that,

$$\left| \det \frac{\partial x}{\partial y}(y, \eta) \right| \geq \tilde{\delta}, \quad y, \eta \in \mathbb{R}^d.$$

Conversely, every transformation χ satisfying $A1$, $A2$ and $A3$ associates with a tame phase function Φ_χ, the correspondence being unique up to a constant (see e.g. [54]).

The following result was proved in [53, Theorem 3.3].

Theorem 4.4.4 *Let $\Phi(x, \eta)$ be a tame phase function, and let χ be the corresponding canonical transformation. Let $a(x, \eta)$ be a function in $C_b^\infty(\mathbb{R}^{2d})$. The Fourier integral operator*

$$Tf(x) = \int_{\mathbb{R}^d} e^{2\pi i \Phi(x, \eta)} a(x, \eta) \hat{f}(\eta) \, d\eta$$

belongs to $FIO(\chi)$. Moreover, for every $m \in \mathbb{N}$ there exists $m' \in \mathbb{N}$ such that

$$\|T\|_{m, \chi} \leq C \|a\|_{m'}$$

for some constant C depending only on m, the dimension d, upper bounds for a certain number of the derivatives of Φ in (4.20) and the lower bound constant $\tilde{\delta}$ in (4.21) (recall that $\|a\|_{m'} = \sup_{|\alpha| + |\beta| \leq m'} \|\partial_x^\alpha \partial_\eta^\beta a\|_{L^\infty}$).

4.4.2 Generalized Metaplectic Operators

In this section we consider the special case where χ is linear, hence it can be identified with a symplectic matrix S. In addition to the classes $FIO(S, v_s)$ that have been studied in the previous section (we just set $S = \chi$), we now introduce the (unweighted) classes $FIO'(S)$ characterized by integrable control functions, following [54, 55]. To be precise, given $S \in \text{Sp}(d, \mathbb{R})$ and $g \in \mathcal{S}(\mathbb{R}^d) \setminus \{0\}$, we say that a linear operator $T: S(\mathbb{R}^d) \to S'(\mathbb{R}^d)$ is in the class $FIO'(S)$ of *generalized metaplectic operators* if there exists $H \in L^1(\mathbb{R}^{2d})$ such that

$$|\langle T\pi(z)g, \pi(w)g \rangle| \leq H(w - Sz), \quad w, z \in \mathbb{R}^{2d}. \tag{4.22}$$

The choice of the name "generalized metaplectic operators" for members of $FIO'(S)$ stems from the fact that the defining property (4.22) generalizes the estimate in Theorem 4.3.13 for metaplectic operators. In particular, note that $\mu(S) \in FIO'(S)$.

The definition of $FIO'(S)$ does not depend on the choice of $g \in S(\mathbb{R}^d) \setminus \{0\}$. In fact, careful inspection of the proof of [55, Proposition 3.1] reveals that the class of admissible windows for $FIO'(S)$ may be extended to $M^1(\mathbb{R}^d)$, the estimate (4.22) being thus equivalent to its *polarized* version with arbitrary windows $g, \gamma \in M^1(\mathbb{R}^d)$, namely

$$|\langle A\pi(z)g, \pi(w)\gamma \rangle| \leq H(w - Sz), \quad w, z \in \mathbb{R}^{2d}.$$

Sparsity of the Gabor matrix of a generalized metaplectic operator associates with non-trivial algebraic properties for $FIO'(S)$, in the spirit of Theorem 4.2.5. To this aim, we need the following preliminary result.

Lemma 4.4.5 *Let X denote either $M_{0,s}^{\infty}(\mathbb{R}^{2d})$, $s \geq 0$, or $M^{\infty,1}(\mathbb{R}^{2d})$. Let $\sigma \in X$ and A, B, C be real $d \times d$ matrices with B invertible, and set*

$$\Phi(x, y) = \frac{1}{2}Ax \cdot x + Bx \cdot y + \frac{1}{2}Cy \cdot y.$$

There exists a unique symbol $\widetilde{\sigma} \in X$ such that, for any $f \in S(\mathbb{R}^d)$:

$$\sigma^w \int_{\mathbb{R}^d} e^{2\pi i \Phi(x,y)} f(y)dy = \int_{\mathbb{R}^d} e^{2\pi i \Phi(x,y)} \widetilde{\sigma}(x, y) f(y)dy.$$

Furthermore, the map $\sigma \mapsto \widetilde{\sigma}$ is an automorphism of X.

Proof Here we give only a sketch of the proof, we refer to [55, Proposition 5.2], where the case $X = M^{\infty,1}(\mathbb{R}^{2d})$ is treated in detail. An explicit computation shows that $\widetilde{\sigma}$ is directly derived from σ as follows:

$$\widetilde{\sigma} = \mathcal{U}_2 \mathcal{U} \mathcal{U}_1 \sigma,$$

where $\mathcal{U}, \mathcal{U}_1, \mathcal{U}_2$ are the mappings

$$\mathcal{U}_1 \sigma(x, y) = \sigma(x, y + Ax), \quad \mathcal{U}_2 \sigma(x, y) = \sigma(x, B^{\top} y), \quad \widehat{\mathcal{U}\sigma}(\xi, \eta) = e^{\pi i \xi \cdot \eta} \widehat{\sigma}(\xi, \eta).$$

$\mathcal{U}_1$ and $\mathcal{U}_2$ are automorphisms of X, because modulation spaces are easily seen to be invariant with respect to linear change of variables. For what concerns $\mathcal{U}$, a by-product of the proof of [121, Corollary 14.5.5] is that any modulation space $M_{0,s}^{p,q}(\mathbb{R}^{2d})$ is invariant under the action of $\mathcal{U}$. Alternatively, one can obtain the boundedness of the Fourier multiplier $\mathcal{U}$ (and its inverse) from Proposition 3.6.9 since $m(x,\xi) = e^{-\pi i x \cdot \xi}$ belongs to $W^{1,\infty}(\mathbb{R}^{2d})$, cf. [64, Proposition 3.2]. $\square$

Theorem 4.4.6 *Let* $S, S_1, S_2 \in \mathrm{Sp}(d, \mathbb{R})$.

(i) *An operator* $T \in FIO'(S)$ *is bounded on* $M^p(\mathbb{R}^d)$ *for any* $1 \le p \le \infty$.

(ii) *If* $T_1 \in FIO'(S_1)$ *and* $T_2 \in FIO'(S_2)$, *then* $T_1 T_2 \in FIO'(S_1 S_2)$.

(iii) *If* $T \in FIO'(S)$ *is invertible on* $L^2(\mathbb{R}^d)$ *then* $T^{-1} \in FIO'(S^{-1})$.

(iv) *Let* $T \colon \mathcal{S}(\mathbb{R}^d) \to \mathcal{S}'(\mathbb{R}^d)$ *be a linear continuous operator.* $T \in FIO'(S)$ *if and only if there exist* $\sigma_1, \sigma_2 \in M^{\infty,1}(\mathbb{R}^{2d})$ *such that*

$$T = \sigma_1^{\mathrm{w}} \mu(S) = \mu(S) \sigma_2^{\mathrm{w}}.$$

In particular, $\sigma_2 = \sigma_1 \circ S$.

(v) *Let* $S = \begin{bmatrix} A & B \\ C & D \end{bmatrix} \in \mathrm{Sp}_0(d, \mathbb{R})$—*namely* $\det B \ne 0$. *A linear continuous operator* $T \colon \mathcal{S}(\mathbb{R}^d) \to \mathcal{S}'(\mathbb{R}^d)$ *belongs to* $FIO'(S)$ *if and only if there exists* $\sigma \in M^{\infty,1}(\mathbb{R}^{2d})$ *such that it can be represented as an integral operator, namely*

$$Tf(x) = \int_{\mathbb{R}^d} e^{2\pi i \Phi_S(x,y)} \sigma(x,y) f(y) dy,$$

where Φ_S *is the generating function in* (4.4).

The same claims are true for $FIO(S, v_s)$, $s \ge 0$, *provided that* $M^{\infty,1}(\mathbb{R}^{2d})$ *is replaced by* $M_{0,s}^{\infty}(\mathbb{R}^{2d})$ *and* $M^p(\mathbb{R}^d)$ *by* $M_{v_s}^p(\mathbb{R}^d)$.

Proof We provide the proofs for the claims concerning the classes $FIO'(S)$. The case of $FIO(S, v_s)$ follows by analogous arguments with suitable modifications.

As in the proof of Theorem 4.2.4, we set the problem at the level of phase space by means of the Gabor matrix of T, with respect to a window $g \in \mathcal{S}(\mathbb{R}^d)$, $\|g\|_{L^2} = 1$, namely

$$K_T(w, z) = \langle T\pi(z)g, \pi(w)g \rangle.$$

Recall that K_T is the integral kernel of the operator $\widetilde{T} = V_g T V_g^*$—cf. Sect. 4.1.

(i) Boundedness of T on $M^p(\mathbb{R}^d)$ follows from boundedness of $\widetilde{T}$ on $L^p(\mathbb{R}^{2d})$, $1 \le p \le \infty$. Using the defining property of $FIO'(S)$ we thus have

$$|\widetilde{T}F(w)| = \left| \int_{\mathbb{R}^{2d}} K_T(w, z) F(z) dz \right|$$

$$\le \int_{\mathbb{R}^{2d}} |F(z)| H(w - Sz) dz$$

$$= |F| * (H \circ S)(S^{-1}w).$$

Clearly, if $H \in L^1(\mathbb{R}^{2d})$ then $H \circ S \in L^1(\mathbb{R}^{2d})$, hence

$$|F| * (H \circ S) \in L^p(\mathbb{R}^{2d}) * L^1(\mathbb{R}^{2d}) \subset L^p(\mathbb{R}^{2d}),$$

and the claim follows.

(ii) Note that

$$\widetilde{T_1 T_2} = V_g T_1 T_2 V_g^* = (V_g T_1 V_g^*)(V_g T_2 V_g^*) = \widetilde{T}_1 \widetilde{T}_2.$$

Let $H_i \in L^1(\mathbb{R}^{2d})$ be the controlling function for the Gabor matrix of T_i, $i = 1, 2$. Therefore,

$$|\langle T_1 T_2 \pi(z)g, \pi(w)g \rangle| \leq \int_{\mathbb{R}^{2d}} |K_{T_1}(w, u)||K_{T_2}(u, z)|du$$

$$\leq \int_{\mathbb{R}^{2d}} H_1(w - S_1 u) H_2(u - S_2 z)du$$

$$= \int_{\mathbb{R}^{2d}} (H_1 \circ S_1)(S_1^{-1} w - u) H_2(u - S_2 z)du$$

$$= ((H_1 \circ S_1) * H_2) \circ S_1^{-1})(w - S_1 S_2 z).$$

The conclusion follows again from the invariance of L^1 under dilation by invertible matrices and the convolution property $L^1 * L^1 \subset L^1$.

(iii) First, we show that $T^* \in FIO'(S^{-1})$. Indeed,

$$|\langle T^* \pi(z)g, \pi(w)g \rangle| = |\langle \pi(z)g, T(\pi(w)g) \rangle|$$

$$\leq H(z - Sw)$$

$$= (H \circ S)(S^{-1} z - w)$$

$$= (H \circ S)(-(w - S^{-1} z)).$$

In view of the invariance of L^1 under reflection and the previous item we infer that the operator $P := T^* T$ belongs to $FIO'(I)$, hence there exists $H \in L^1(\mathbb{R}^{2d})$ such that

$$K_P(w, z) = |\langle P\pi(z)g, \pi(w)g \rangle| \leq H(w - z).$$

According to the characterization in Theorem 4.2.5 this estimate implies that $P = \sigma^w$ for some $\sigma \in M^{\infty,1}(\mathbb{R}^{2d})$. The Wiener property stated in Theorem 4.2.4 allows us to conclude that P^{-1} is a Weyl operator with symbol in the Sjöstrand class too, hence $P^{-1} \in FIO'(I)$. Using the algebra property of the previous item, we conclude $T^{-1} = P^{-1} T^* \in FIO'(S^{-1})$ as claimed.

(iv) We first prove that $T \in FIO'(S)$ implies the factorization $T = \sigma_1^w \mu(S)$. Note that $\mu(S)^{-1} = \mu(S^{-1}) \in FIO'(S^{-1})$, hence the algebra property for the classes FIO' proved above yields $T\mu(S^{-1}) \in FIO'(I)$. Arguing as in the previous item, we deduce that $T\mu(S^{-1}) = \sigma_1^w$ for some $\sigma_1 \in M^{\infty,1}(\mathbb{R}^{2d})$, that is the claim. Moreover, using the symplectic covariance of the Weyl calculus (4.16) we have $T = \mu(S)\mu(S)^{-1}\sigma_1^w \mu(S) = \mu(S)(\sigma_1 \circ S)^w$.

Conversely, consider $T = \mu(S)\sigma_1^w$ with $\sigma_1 \in M^{\infty,1}(\mathbb{R}^{2d})$. Since $\mu(S) \in FIO'(S)$ and $\sigma_1^w \in FIO'(I)$ as already showed before, we conclude that $T \in FIO'(S)$ using the algebra property.

(v) Assume that $T \in FIO'(S)$. As a consequence of the previous item we have that the factorization $T = \sigma_1^w \mu(S)$ holds for some $\sigma_1 \in M^{\infty,1}(\mathbb{R}^{2d})$. The claimed result is thus obtained by combining the representation formula (4.9) for $\mu(S)$ with Lemma 4.4.5.

Conversely, assume that T is an integral operator as in the statement. Using again Lemma 4.4.5 and the representation formula (4.9) we recognize that the factorization $T = \sigma_0^w \mu(S)$ holds for a suitable $\sigma_0 \in M^{\infty,1}(\mathbb{R}^{2d})$, hence $T \in FIO'(S)$.

$\square$

Remark 4.4.7 We highlight that similar arguments involving Theorem 4.3.7 allow us to prove that if $S = \begin{bmatrix} A & B \\ C & D \end{bmatrix} \in \mathrm{Sp}(d, \mathbb{R})$ is such that $\det A \neq 0$ then $T \in FIO'(S)$ if and only if it can be represented as a Fourier integral operator, namely

$$Tf(x) = \int_{\mathbb{R}^d} e^{2\pi i \Phi_{SJ^{-1}(x,y)}} \sigma(x, y) \hat{f}(y) dy, \quad f \in S(\mathbb{R}^d),$$

for some $\sigma \in M^{\infty,1}(\mathbb{R}^{2d})$. This provides some justification for the name FIO used for the class of operators under our consideration.

Remark 4.4.8 In view of Theorem 4.4.6 and arguing as we did before for metaplectic operators, we observe that the Gabor matrix $\langle T\pi(z)g, \pi(w)\gamma \rangle$ of a generalized metaplectic operator $T \in FIO(S)$ is well defined in the case where

$$g \in M^p(\mathbb{R}^d), \ \gamma \in M^q(\mathbb{R}^d), \quad \frac{1}{p} + \frac{1}{q} \geq 1.$$

4.4.3 Oscillatory Integral Operators with Rough Amplitude

Let T be an oscillatory integral operator of the form

$$Tf(x) = \int_{\mathbb{R}^d} e^{2\pi i \lambda \Phi(x,y)} a(x, y) f(y) \, dy, \quad \lambda \geq \lambda_0 > 0. \tag{4.23}$$

Several conditions on the phase function $\Phi(x, y)$ and the amplitude $a(x, y)$ are known for A to be bounded on $L^2(\mathbb{R}^d)$; see e.g. [47, 197, 220] and the references therein. Here we recall the following result from [27, Corollary 1].

Proposition 4.4.9 *Let T be the operator in (4.23), and suppose Φ real-valued with $\partial^\alpha \Phi \in H^r_{\mathrm{ul}}(\mathbb{R}^{2d})$ for $|\alpha| = 2$, and $a \in H^r_{\mathrm{ul}}(\mathbb{R}^{2d})$ for some $r > d$. Assume, moreover, that*

$$\left| \det \frac{\partial^2 \Phi}{\partial x \partial y}(x, y) \right| \geq \tilde{\delta}, \qquad x, y \in \mathbb{R}^d,$$

for some $\tilde{\delta} > 0$.

Then T, initially defined on Schwartz functions, extends to a bounded operator on $L^2(\mathbb{R}^d)$. In fact, there exists a constant $C > 0$ independent of $\lambda \geq \lambda_0$, $\tilde{\delta}$, Φ and a such that

$$\|T\|_{L^2 \to L^2} \leq C\tilde{\delta}^{-1}\lambda^{-d/2} \exp(C\|D^2\Phi\|_{H^r_{\mathrm{ul}}(\mathbb{R}^{2d})})\|a\|_{H^r_{\mathrm{ul}}(\mathbb{R}^{2d})},$$

where $D^2\Phi$ denotes the Hessian matrix of Φ.

We also recall from [28, Theorem 2.1] a boundedness result for oscillatory integral operators in terms of the $M^{\infty,1}$ norm of the amplitude.

Lemma 4.4.10 *Consider the oscillatory integral operator*

$$Tf(x) = \int_{\mathbb{R}^d} e^{i\frac{|x-y|^2}{2}} a(x, y) f(y) \, dy, \qquad f \in \mathcal{S}(\mathbb{R}^d),$$

with $a \in M^{\infty,1}(\mathbb{R}^{2d})$. Then T extends to a bounded operator in $L^2(\mathbb{R}^d)$ and there exists a constant $C > 0$, depending only on d, such that

$$\|T\|_{L^2 \to L^2} \leq C\|a\|_{M^{\infty,1}}.$$

Note that by expanding the phase $|x - y|^2/2$ one could deduce this result from known boundedness results for Kohn-Nirenberg pseudodifferential operators [121, Corollary 14.5.5] and the Parseval formula for the Fourier transform. It is also a consequence of Theorem 4.4.4.

4.5 Complements

4.5.1 Weyl Operators and Narrow Convergence

The result in Theorem 4.2.2 extends to weighted spaces, as well as Wiener amalgam spaces, as detailed below; in this connection, see also [222, Theorem 2.13].

Theorem 4.5.1 *Let $1 \leq p, q \leq \infty$, $\gamma \geq 0$ and $r, s \in \mathbb{R}$ be such that $|r| + |s| \leq \gamma$; denote by X any of the spaces $M_{r,s}^{p,q}(\mathbb{R}^d)$ or $W_{r,s}^{p,q}(\mathbb{R}^d)$. If $\sigma \in M_{0,\gamma}^{\infty,1}(\mathbb{R}^{2d})$ then the Weyl operator σ^w is bounded on X.*

Proof The case $X = M_{r,s}^{p,q}(\mathbb{R}^d)$ follows from [122, Theorem 3.2] (it is a generalization of Theorem 4.2.5 to weighted Sjöstrand classes) and convolution relations for weighted Lebesgue spaces, in particular $L_m^{p,q}(\mathbb{R}^{2d}) * L_v^1(\mathbb{R}^{2d}) \subset L_m^{p,q}(\mathbb{R}^{2d})$, cf. [121, Proposition 11.1.3]. For the case $X = W_{r,s}^{p,q}(\mathbb{R}^d)$ we need a special variant of the symplectic covariance of Weyl calculus (4.16), namely

$$\mathcal{F}\sigma^w = \sigma_{J^{-1}}^w \mathcal{F}, \qquad \sigma \in \mathcal{S}'(\mathbb{R}^{2d}),$$

where $\sigma_{J^{-1}} = \sigma \circ J^{-1}$—recall that $\mathcal{F} = c\mu(J)$ for some $c \in \mathbb{C}$, $|c| = 1$. In view of this property, consider the following diagram:

$$\begin{array}{ccc} M_{r,s}^{p,q}(\mathbb{R}^d) & \xrightarrow{\sigma_J^w} & M_{r,s}^{p,q}(\mathbb{R}^d) \\ \Big\uparrow{\mathcal{F}^{-1}} & & \Big\downarrow{\mathcal{F}} \\ W_{r,s}^{p,q}(\mathbb{R}^d) & \xrightarrow{\sigma^w} & W_{r,s}^{p,q}(\mathbb{R}^d) \end{array}$$

It is easy to prove that if $\sigma \in M_{0,\gamma}^{\infty,1}(\mathbb{R}^{2d})$ then $\sigma_J \in M_{0,\gamma}^{\infty,1}(\mathbb{R}^{2d})$ too (cf. for instance the proof of [66, Lemma 5.2]), hence the preceding case implies that σ_J^w is bounded on $M_{r,s}^{p,q}(\mathbb{R}^d)$ for any $1 \leq p, q \leq \infty$ and $r, s \in \mathbb{R}$ such that $|r| + |s| \leq \gamma$. $\qquad\square$

There is also a variant for families of symbols in weighted Sjöstrand classes, provided that the dependence from the parameter is continuous for the narrow convergence—cf. Sect. 3.6.5. In the following statements we denote by $\mathcal{M}_{r,s}^{p,q}(\mathbb{R}^d)$ and $\mathcal{W}_{r,s}^{p,q}(\mathbb{R}^d)$ the closure of $\mathcal{S}(\mathbb{R}^d)$ under the norms of $M_{r,s}^{p,q}$ and $W_{r,s}^{p,q}$ respectively. Such spaces do coincide with standard modulation and amalgam spaces except for $p = \infty$ or $q = \infty$—in such cases the latter are strictly larger. This caution is necessary in order to avoid technical difficulties arising in such situation [217, 218].

Theorem 4.5.2 *For any $1 \leq p, q \leq \infty$ and $\gamma \geq 0$, $r, s \in \mathbb{R}$ such that $|r| + |s| \leq \gamma$, let X denote either $\mathcal{M}_{r,s}^{p,q}(\mathbb{R}^d)$ or $\mathcal{W}_{r,s}^{p,q}(\mathbb{R}^d)$. If $\Omega \ni v \mapsto \sigma_v \in M_{0,\gamma}^{\infty,1}(\mathbb{R}^{2d})$ is continuous for the narrow convergence then the corresponding map of operators $v \mapsto \sigma_v^w$ is strongly continuous on X.*

Proof The proof for $X = \mathcal{M}_{r,s}^{p,q}(\mathbb{R}^d)$ can be found in [58, Proposition 3]. For the strong continuity on $X = \mathcal{W}_{r,s}^{p,q}(\mathbb{R}^d)$ we reduce to the latter case by the same arguments as in the proof of Proposition 4.5.1, which imply that $\sigma_v^w u = \mathcal{F}(\sigma_v)_J^w \mathcal{F}^{-1} u$ for $u \in \mathcal{W}_{r,s}^{p,q}(\mathbb{R}^d)$. The claimed result easily follows from the continuity of the map $v \mapsto (\sigma_v)_J^w \mathcal{F}^{-1} u$ on $\mathcal{M}_{r,s}^{p,q}(\mathbb{R}^d)$. $\qquad\square$

4.5.2 General Quantization Rules

Recall that in Sect. 3.6.2 we introduced the Cohen class of time-frequency distributions, parametrized by a kernel $\theta \in S'(\mathbb{R}^{2d})$, that is

$$Q_\theta(f, g) := W(f, g) * \theta, \qquad f, g \in S(\mathbb{R}^d).$$

For every time-frequency representation in Cohen's class one can naturally introduce a quantization rule in analogy to the Weyl quantization (1.5), namely,

$$\langle \mathrm{op}_\theta(\sigma) f, g \rangle = \langle \sigma, Q_\theta(g, f) \rangle, \quad f, g \in S(\mathbb{R}^d), \qquad (4.24)$$

whenever the expressions make sense [46, 121].

Although the new operator $\mathrm{op}_\theta(\sigma)$ is ultimately a Weyl operator with the modified symbol $\sigma * \theta^*$ (whenever defined in $S'(\mathbb{R}^d)$—recall that $\theta^*(z) = \overline{\theta(-z)}$), the machinery given by definition (4.24) adds a new flavour to the analysis of pseudodifferential operators, as they can be interpreted as deformations of the standard Weyl transform. For example, a first important variation of the Wigner transform are the τ-*Wigner transforms* [23, 24] as defined in Sect. 3.6.2. Recall that they are parametrized by a real number τ and are defined as

$$W_\tau(f, g)(x, \xi) = \int_{\mathbb{R}^d} e^{-2\pi i y \cdot \xi} f(x + \tau y) \overline{g(x - (1 - \tau)y)} \, dy, \quad f, g \in S(\mathbb{R}^d).$$

Such distributions belong to Cohen's class, their kernel $\theta_\tau \in S'(\mathbb{R}^{2d})$ given by

$$\widehat{\theta_\tau}(x, \xi) = e^{-2\pi i(\tau - 1/2)x \cdot \xi}, \quad (x, \xi) \in \mathbb{R}^{2d}.$$

The corresponding pseudodifferential calculi are the Shubin τ-operators [207]; according to formula (4.24) these are explicitly given by

$$\mathrm{op}_\tau(\sigma) f(x) = \int_{\mathbb{R}^{2d}} e^{2\pi(x-y)\cdot\xi} \sigma((1 - \tau)x + \tau y, \xi) f(y) dy d\xi. \qquad (4.25)$$

In some sense, the dependence on x and y of the symbol is now amalgamated in an affine combination of the variables. Note that for the parameter $\tau = 1/2$ this is just the Weyl transform, while for $\tau = 0$ we recover the standard Kohn-Nirenberg quantization:

$$\mathrm{op}_{\mathrm{KN}}(\sigma) f = \int_{\mathbb{R}^d} e^{2\pi(x-y)\cdot\xi} \sigma(x, \xi) f(y) dy d\xi.$$

We also highlight that the important Born-Jordan quantization rule [26] also belong to the Cohen class and in fact can be obtained as an integral average over $\tau \in [0, 1]$, see [23, 62, 227] and the monograph [72].

While the Cohen class admittedly comprises more general distributions, τ-distributions are characterized by a precise, explicit control of the deviation from the Wigner distribution thanks to the parameter τ—or even $\mu = \tau - 1/2$. Accordingly, it is interesting to investigate whether some of the most relevant properties of the Wigner distributions and the Weyl operators "survive" the perturbation, that is if they extend (possibly in a weaker form) to τ distributions and the corresponding quantizations. This perspective has been embraced in several papers: for instance, in [65] uniform upper bounds with respect to $\tau \in (0, 1)$ were proved for the τ-Wigner distributions and, by duality, for the operator norm of $\mathrm{op}_\tau(\sigma)$ on modulation and Wiener amalgam spaces for several symbol classes of the same type, while in the article [66] quasi-diagonalization results in the spirit of Theorem 4.2.5 were proved for τ-operators with symbols in $M^{\infty,1}$ or $W^{\infty,1}$.

In general, one could try to fill the gap between generality and controllability by allowing general parametrizations $\tau : \mathbb{R}^d \rightarrow \mathbb{R}^d$ in (4.25). For instance, in the recent paper [78] the authors considered smooth quantizing functions with bounded or unbounded derivatives. While in the nonlinear scenario there is still too much freedom in choosing the function τ, the linear case can be characterized in full generality: this just amounts to replace the scalar parameter τ with a matrix parameter $T \in \mathbb{R}^{d \times d}$. The resulting family of T-*pseudodifferential operators* is

$$\sigma^T f(x) = \int_{\mathbb{R}^{2d}} e^{2\pi i(x-y)\cdot\xi} \sigma((I - T)x + Ty, \xi) f(y) dy d\xi = \langle \sigma, W_T(g, f) \rangle,$$

where the matrix-Wigner distribution $W_T(g, f)$ has been introduced in (3.13).

These are members of the Cohen class in (3.12) with a kernel θ_T given by

$$\theta_T = \mathcal{F}^{-1} \Theta_T \in \mathcal{S}'(\mathbb{R}^{2d}), \quad \Theta_T(u, v) = e^{-2\pi i \xi \cdot (T - I/2)\eta}.$$

The first thorough investigation of such matrix-parametrized distributions and operators—even in a more general form, beyond the Cohen class—is contained in the unpublished Ph.D. thesis [11] of D. Bayer. The topic also appeared occasionally in the literature, cf. for instance [22, 37, 117, 219]. We refer to the recent papers [12, 51] for a more systematic discussion.

4.5.3 The Class $FIO'(S, v_s)$

Along with the classes $FIO(S, v_s)$ and $FIO'(S)$ introduced in Sect. 4.4 before, one may also consider the families $FIO'(S, v_s)$ of generalized metaplectic operators with controlling function in $L_s^1(\mathbb{R}^{2d})$. To be precise, following [55, Definition 1.1], given $S \in \mathrm{Sp}(d, \mathbb{R})$, $s \geq 0$ and $g \in \mathcal{S}(\mathbb{R}^d) \setminus \{0\}$, the class $FIO'(S, v_s)$ consists of all linear continuous operators $A : \mathcal{S}(\mathbb{R}^d) \rightarrow \mathcal{S}'(\mathbb{R}^d)$ such that there exists $H \in L_s^1(\mathbb{R}^{2d})$ which controls the decay of the Gabor matrix of A, namely

$$|\langle A\pi(z)g, \pi(w)g \rangle| \leq H(w - Sz), \quad w, z \in \mathbb{R}^{2d}.$$

Note in particular that the class $FIO'(S)$ studied before coincides with the unweighted setting where $s = 0$.

The results stated in Theorem 4.4.6 for $FIO'(S)$ extend to $FIO'(S, v_s)$ provided that $M^{\infty,1}(\mathbb{R}^{2d})$ is replaced by $M_{0,s}^{\infty,1}(\mathbb{R}^{2d})$ and $M^p(\mathbb{R}^d)$ by $M_{v_r}^p(\mathbb{R}^d)$, $|r| \leq s$.

4.5.4 Finer Aspects of Gabor Wave Packet Analysis

Roughly speaking, the *leitmotiv* of this chapter has been to exploit the phase-space characterizations of several classes of operators in order to deduce results of various kinds (e.g., boundedness, composition, explicit representations). This program has been carried out concretely by requiring some control on the decay of the Gabor matrix of pseudodifferential operators, Fourier integral operators and propagators associated with Cauchy problems for Schrödinger-type evolution equations. We recommend the recent monograph [50] for a detailed and systematic account, and the papers [36, 68, 130, 210, 214] for other applications of wave packet analysis.

In this section we wish to emphasize how the analysis with Gabor wave packets is able to provide interesting results even at deeper levels. In particular, we show how to recapture and explain the occurrence of peculiar phenomena in Gabor matrices, such as dispersion, spreading and sparsity.

As a motivation, let us consider the Schrödinger propagator $U(t) = e^{-2\pi i t Q^w}$ associated with a quadratic Hamiltonian Q as in Sect. 4.3.3. We already remarked that $U(t)$ is a metaplectic operator, in particular $U(t) = \mu(S_t)$, where $\mathbb{R} \ni t \mapsto S_t \in \mathrm{Sp}(d, \mathbb{R})$ coincides with the corresponding classical Hamiltonian flow. As such, from Theorem 4.3.13 we infer that the Gabor matrix is well-organized: precisely, for any $t \in \mathbb{R}$ and $g \in \mathcal{S}(\mathbb{R}^d) \setminus \{0\}$ we have

$$|\langle U(t)\pi(z)g, \pi(w)g \rangle| \lesssim (1 + |w - S_t z|)^{-N}, \quad w, z \in \mathbb{R}^{2d}. \tag{4.26}$$

It is then clear that the propagator $U(t)$ evolves Gabor wave packets under the influence of the companion classical dynamics in phase space, i.e., approximately along the graph of the classical flow S_t. This phenomenon can be suggestively viewed as a phase-space instance of the correspondence principle of quantum mechanics.

Whenever concerned with wave propagation dynamics one should also take into account the distinctive and unavoidable trait of diffraction. In the situation under current examination it does take the form of a well-known phenomenon, which is the *spreading* of wave packets [75, Section 31].

There is one more aspect to reckon with, namely the *dispersive* nature of the Schrödinger propagator. For the sake of simplicity, let us consider the free particle propagator $U(t) = e^{i(t/4\pi)\Delta}$. Using the standard dispersive estimates for the

Schrödinger propagator [213] it is not difficult to show that there exists $C > 0$ such that

$$|\langle e^{i(t/4\pi)\Delta}\pi(z)g, \pi(w)g\rangle| \leq C(1 + |t|)^{-d/2}, \quad w, z \in \mathbb{R}^{2d}. \tag{4.27}$$

In the light of the above, it may seem quite disappointing that there is no trace of such issues in quasi-diagonalization estimates as (4.26). This question has been addressed in the recent article [67], where refined estimates for the Gabor matrix of a metaplectic operator were provided—in particular, sparsity, spreading and dispersive phenomena are simultaneously represented.

The spreading of wave packets under the action of a metaplectic operator $\mu(S)$ is connected with the *singular values* of $S \in \mathrm{Sp}(d, \mathbb{R})$ [40], which occur in couples (σ, σ^{-1}) of positive numbers. We fix the ordering by labelling the largest d singular values in such a way that $\sigma_1 \geq \ldots \geq \sigma_d \geq 1$; moreover we set $\Sigma = \mathrm{diag}(\sigma_1, \ldots, \sigma_d)$ and introduce the matrices

$$D = \begin{bmatrix} \Sigma & O \\ O & \Sigma^{-1} \end{bmatrix}, \quad D' = \begin{bmatrix} \Sigma^{-1} & O \\ O & I \end{bmatrix}, \quad D'' = \begin{bmatrix} I & O \\ O & \Sigma^{-1} \end{bmatrix}.$$

We recall from Proposition 4.3.2 that there exist (non-unique) orthogonal and symplectic matrices U, V such that $S = U^{\top}DV$. In the following, for a given $S \in \mathrm{Sp}(d, \mathbb{R})$ we will denote by (U, V, Σ) such an Euler decomposition of S and by D, D', D'' the above defined related matrices.

We recall two results from [67], involving Gabor wave packets at different regularity levels. The first one concerns rapidly decaying wave packets and is a refinement of Theorem 4.3.13.

Theorem 4.5.3 *For any $g, \gamma \in \mathcal{S}(\mathbb{R}^d)$ and $N > 0$ there exists $C > 0$ such that, for every $S \in \mathrm{Sp}(d, \mathbb{R})$ and any Euler decomposition (U, V, Σ) of S,*

$$|\langle \mu(S)\pi(z)g, \pi(w)\gamma\rangle| \leq C(\det \Sigma)^{-1/2}(1 + |D'U(w - Sz)|)^{-N}, \quad z, w \in \mathbb{R}^{2d}. \tag{4.28}$$

We see that this estimate concurrently encompasses the sparsity, spreading and dispersive phenomena discussed above, which are respectively represented by the quasi-diagonal structure along S, the dilation by $D'U$ and the factor $(\det \Sigma)^{-1/2}$. In passing, we remark that the spreading effect can be somehow distributed after noticing that $D'U(w - Sz) = D'Uw - D''Vz$.

A result in the same spirit holds for wave packets associated with less regular atoms; in particular, we assume that g and γ belong to suitable modulation spaces.

Theorem 4.5.4

(i) Let $1 \le p, q, r \le \infty$ satisfy $1/p + 1/q = 1 + 1/r$. For any $g \in M^p(\mathbb{R}^d)$, $\gamma \in M^q(\mathbb{R}^d)$, $S \in \mathrm{Sp}(d, \mathbb{R})$ and any Euler decomposition (U, V, Σ) of S, there exists $H \in L^r(\mathbb{R}^{2d})$ such that, for any $z, w \in \mathbb{R}^{2d}$,

$$|\langle \mu(S)\pi(z)g, \pi(w)\gamma \rangle| \le H(D'U(w - Sz)), \tag{4.29}$$

with

$$\|H\|_{L^r} \le (\det \Sigma)^{1/2 - 1/r} \|g\|_{M^p} \|\gamma\|_{M^q}. \tag{4.30}$$

(ii) Let $s > 2d$. For any $g, \gamma \in M^\infty_{v_s}(\mathbb{R}^d)$ there exists $H \in L^\infty_{s-2d}(\mathbb{R}^{2d})$ such that (4.29) holds, with

$$\|H\|_{L^\infty_{s-2d}} \le (\det \Sigma)^{-1/2} \|g\|_{M^\infty_{v_s}(\mathbb{R}^d)} \|\gamma\|_{M^\infty_{v_s}(\mathbb{R}^d)}.$$

Note that the best decay in (4.30) is achieved for Gabor atoms belonging to the Feichtinger algebra $M^1(\mathbb{R}^d)$ ($p = q = r = 1$). We also highlight the inclusion $M^\infty_{v_s}(\mathbb{R}^d) \subset M^1(\mathbb{R}^d)$ for $s > 2d$, which follows directly from the definition.

Finally, a result in the same spirit is provided also for generalized metaplectic operators, introduced in Sect. 4.4.2.

Theorem 4.5.5 *Let* $1 \le p, q, r \le \infty$ *satisfy* $1/p + 1/q = 1 + 1/r$. *Consider* $S \in \mathrm{Sp}(d, \mathbb{R})$ *with an Euler decomposition* (U, V, Σ), $a \in M^{\infty,1}(\mathbb{R}^{2d})$ *so that* $A := a^w\mu(S) \in FIO'(S)$, *cf. Theorem 4.4.6. For any* $g \in M^p(\mathbb{R}^d)$, $\gamma \in M^q(\mathbb{R}^d)$ *there exists* $H \in L^r(\mathbb{R}^{2d})$ *such that, for any* $z, w \in \mathbb{R}^{2d}$,

$$|\langle A\pi(z)g, \pi(w)\gamma \rangle| \le H(D'U(w - Sz)),$$

with

$$\|H\|_{L^r} \le (\det \Sigma)^{1/2 - 1/r} \|a\|_{M^{\infty,1}} \|g\|_{M^p} \|\gamma\|_{M^q}.$$

Applications of these results to the problems of boundedness and propagation of singularities for metaplectic operators are provided in [67]. Here we just confine ourselves to discuss the special case of the free particle propagator $U_0(t) = e^{i(t/4\pi)\Delta}$, in order to better understand the role of the terms in the estimates for the Gabor matrix with a concrete example.

First, a straightforward computation shows that the largest d singular values of the corresponding free classical flow

$$S_t = \begin{bmatrix} I & tI \\ O & I \end{bmatrix}, \quad t \in \mathbb{R},$$

all coincide: $\sigma_j = \sigma(t) = \sqrt{1 + t^2/4} + |t|/2$, $j = 1, \ldots, d$. Note in particular that $\sigma(t)$ is comparable to $1 + |t|$, $t \in \mathbb{R}$, hence for any fixed $t \in \mathbb{R}$ and any Euler decomposition (U_t, V_t, Σ_t) of S_t, the estimate (4.28) reads

$$\left| \langle e^{i(t/4\pi)\Delta} \pi(z)g, \pi(w)\gamma \rangle \right| \le C(1+|t|)^{-d/2}(1 + |D_t' U_t(w - S_t z)|)^{-N}, \quad w, z \in \mathbb{R}^{2d}.$$

We see that the features of both (4.26) and (4.27) are now represented, whereas the spreading phenomenon manifests itself as a dilation by the matrix $D_t' U_t$.

To be concrete, let us consider $t \ge 0$; an example of Euler decomposition (U_t, V_t, Σ_t) of S_t is given by

$$U_t = (1 + \sigma(t)^2)^{-1/2} \begin{bmatrix} \sigma(t)I & I \\ -I & \sigma(t)I \end{bmatrix}, \quad V_t = (1 + \sigma(t)^2)^{-1/2} \begin{bmatrix} I & \sigma(t)I \\ -\sigma(t)I & I \end{bmatrix},$$

and the spreading phenomenon corresponds to the dilation by

$$D_t' U_t = (1 + \sigma(t)^2)^{-1/2} \begin{bmatrix} I & \sigma(t)^{-1}I \\ -I & \sigma(t)I \end{bmatrix}.$$

In an attempt to elucidate the puzzling structure of this matrix we consider a toy example in dimension $d = 1$. Let $z = 0$ for simplicity and assume that the atom g is concentrated on the unit ball $B \subset \mathbb{R}^2$ in the time-frequency plane. In view of (4.28) we are led to consider the set

$$(D_t' U_t)^{-1}(B) = \{(x, \xi) \in \mathbb{R}^2 : |D_t' U_t(x, \xi)| \le 1\}.$$

It is not difficult to realize that the effect of $D_t' U_t$ on Q ultimately amounts to a horizontal stretch by a factor of approximately $\sigma(t)$. This is definitely consistent with the expected spreading of wave packets—see for instance [52, Figures 1-8] for supporting evidence. We wish to emphasize once more that the original estimate (4.26) only captures the coarse-scale classical dynamics, but is completely blind to subtler phase-space phenomena.

Chapter 5
Semiclassical Gabor Analysis

We can safely affirm that all the topics introduced so far are motivated by problems in quantum mechanics, some more and some less. Semiclassical analysis is another example of a branch of mathematics whose original inspiration comes from the problem of reproducing the results of classical mechanics by considering a suitable limit regime of quantum mechanics, as prescribed by Bohr's correspondence principle [25].

The general framework of semiclassical analysis is characterized by the presence of a small parameter $0 < \hbar \leq 1$ and the interest in the corresponding asymptotics $\hbar \to 0$ [242]. Inspired by the reduced Planck constant, one also introduces a companion parameter $h \in (0, 2\pi]$, that is related to $\hbar$ by the formula $\hbar = h/(2\pi)$. We stress that h denotes a dimensionless parameter which should not be systematically identified with the true Planck constant—the latter being just a physical motivation for this mathematical scenario.

The purpose of this chapter is to recast in the spirit of semiclassical analysis some aspects of Gabor analysis that have been discussed before. This is a necessary step in view of the applications to path integrals, but it is also an interesting exercise in general; we refer to the article [61] for a more systematic treatment. The key observation is that

$$e^{ith\Delta} = D_{h^{-1/2}} e^{it\Delta} D_{h^{1/2}},$$

where $D_\lambda f(x) = \lambda^{d/2} f(\lambda x)$ for $\lambda > 0$, as one sees by a rescaling argument in the Schrödinger equation for the free particle or even from the explicit formula for the corresponding propagator. This formula motivates the definiton of the semiclassical versions of known function spaces by conjugation with the (unitary) operator $D_{h^{1/2}}$.

© The Author(s), under exclusive license to Springer Nature Switzerland AG 2022   109
F. Nicola, S. I. Trapasso, *Wave Packet Analysis of Feynman Path Integrals*,
Lecture Notes in Mathematics 2305, https://doi.org/10.1007/978-3-031-06186-8_5

5.1   Semiclassical Transforms and Function Spaces

Let us commence by recalling the notation for the dilation of a function: for $f \colon \mathbb{R}^d \to \mathbb{C}$ and $\lambda \in \mathbb{R}$ we set

$$\delta_\lambda f(t) := f(\lambda t), \quad D_\lambda f(t) := |\lambda|^{d/2} f(\lambda t).$$

It may be useful to highlight that D_λ is a metaplectic operator; precisely, using the notation of Sect. 4.3, we have, for some $c \in \mathbb{C}$, $|c| = 1$,

$$D_\lambda = c\mu(M_{\lambda^{-1}}), \quad M_{\lambda^{-1}} \equiv M_{\lambda^{-1}I} = \begin{bmatrix} \lambda^{-1}I & O \\ O & \lambda I \end{bmatrix}.$$

The semiclassical time-frequency shift acting on f is defined by

$$\pi^\hbar(x, \xi) f(y) = e^{\frac{i}{\hbar}\xi \cdot y} f(y - x), \quad (x, \xi) \in \mathbb{R}^{2d}.$$

Note in particular the relation with the standard time-frequency shift used insofar:

$$\pi(x, \xi) = \pi^\hbar(x, 2\pi\hbar\xi).$$

We also introduce the $\hbar$-dependent Fourier transform of $f \in \mathcal{S}(\mathbb{R}^d)$ by

$$\mathcal{F}^\hbar f(\xi) = (2\pi\hbar)^{-d} \int_{\mathbb{R}^d} e^{-\frac{i}{\hbar}x \cdot \xi} f(x) dx, \quad \xi \in \mathbb{R}^d.$$

Let $1 \le p, q \le \infty$, and fix $g \in \mathcal{S}(\mathbb{R}^d) \setminus \{0\}$. The semiclassical modulation space $M_\hbar^{p,q}(\mathbb{R}^d)$ is the set of all temperate distributions $f \in \mathcal{S}'(\mathbb{R}^d)$ such that

$$\|f\|_{M_\hbar^{p,q}} := \|D_{h^{1/2}} f\|_{M_m^{p,q}} < \infty.$$

We also set $M_\hbar^p(\mathbb{R}^d) = M_\hbar^{p,p}(\mathbb{R}^d)$ for $1 \le p \le \infty$. We have that $M_\hbar^{p,q}(\mathbb{R}^d) = M^{p,q}(\mathbb{R}^d)$ as vector spaces, but the norm $\| \cdot \|_{M_\hbar^{p,q}}$ will be crucial when dealing with uniform estimates with respect to $\hbar$.

It is not difficult to relate the properties enjoyed by standard modulation spaces to those satisfied by their semiclassical versions; for instance, if $1 \le p, q < \infty$ we have

$$(M_\hbar^{p,q}(\mathbb{R}^d))' \simeq M_{1/\hbar}^{p',q'}(\mathbb{R}^d).$$

In order to be more concrete we compute the STFT of $D_{h^{1/2}} f$ with a window $g \in S(\mathbb{R}^d) \setminus \{0\}$. A straightforward computation yields

$$V_g(D_{h^{1/2}} f)(x, \xi) = \langle D_{h^{1/2}} f, \pi(x, \xi) g \rangle$$

$$= \langle f, D_{h^{-1/2}} \pi^\hbar(x, h\xi) g \rangle$$

$$= \langle f, \pi^\hbar(h^{1/2} x, h^{1/2} \xi) D_{h^{-1/2}} g \rangle.$$

Therefore, it may be convenient to introduce the semiclassical Gabor transform of $f \in L^2(\mathbb{R}^d)$ with window $g \in L^2(\mathbb{R}^d) \setminus \{0\}$:

$$V_g^\hbar f(z) := \langle f, \pi^\hbar(h^{1/2} z) D_{h^{-1/2}} g \rangle, \quad z \in \mathbb{R}^{2d}.$$

Note that we retrieve the usual setting in the case where $h = 1$. Hence, for some $g \in S(\mathbb{R}^d) \setminus \{0\}$ we can write $\|f\|_{M_\hbar^{p,q}} = \|V_g^\hbar f\|_{L^{p,q}}$.

Similarly we introduce the (cross-)$\hbar$-Wigner transform of $f, g \in L^2(\mathbb{R}^d)$ by

$$W^\hbar(f, g)(x, \xi) := (2\pi\hbar)^{-d} \int_{\mathbb{R}^d} e^{-\frac{i}{\hbar} y \cdot \xi} f\left(x + \frac{y}{2}\right) \overline{g\left(x - \frac{y}{2}\right)} dy.$$

5.1.1 Sobolev Spaces and Embeddings

We introduce for future reference a scale of semiclassical L^p-based Sobolev spaces. For $1 < p < \infty$ and $r \in \mathbb{R}$ define

$$H_\hbar^{r,p}(\mathbb{R}^d) = \{f \in S'(\mathbb{R}^d) : \|f\|_{H_\hbar^{r,p}} = \left\| (1 - h\Delta)^{r/2} f \right\|_{L^p} < \infty\}.$$

Again, $H_\hbar^{r,p}$ coincides with the usual Sobolev space as far as the vector space structure is concerned, while the norm is rescaled at the Planck scale.

The following embedding results will play a key role in the following.

Theorem 5.1.1 *Let* $1 < p < \infty$ *and* $r = 2d|1/2 - 1/p|$. *There exists a constant* $C > 0$ *such that*

$$\|f\|_{L^p} \leq C h^{d(1/p - 1/2)/2} \|f\|_{M_\hbar^p}, \quad \|f\|_{M_\hbar^p} \leq C h^{d(1/2 - 1/p)/2} \|f\|_{H_\hbar^{r,p}}, \quad 1 < p \leq 2,$$
$$(5.1)$$

as well as

$$\|f\|_{M_\hbar^p} \leq C h^{d(1/2 - 1/p)/2} \|f\|_{L^p}, \quad \|f\|_{H_\hbar^{-r,p}} \leq C h^{d(1/p - 1/2)/2} \|f\|_{M_\hbar^p}, \quad 2 \leq p < \infty.$$
$$(5.2)$$

Proof By (3.8) we have, for $1 < p \leq 2$ and $r = 2d|1/p - 1/2|$,

$$\|f\|_{L^p} \leq C\|f\|_{M^p}, \quad \|f\|_{M^p} \leq C\|f\|_{H^{r,p}}.$$

Now we replace f by $D_{h^{1/2}} f$ and we obtain (5.1), because

$$\|D_{h^{1/2}} f\|_{L^p} = h^{d(1/2-1/p)/2}\|f\|_{L^p} \quad \text{and} \quad \|D_{h^{1/2}} f\|_{H^{r,p}} = h^{d(1/2-1/p)/2}\|f\|_{H_h^{r,p}}.$$

Similarly one deduces (5.2).   □

5.2 Semiclassical Quantization, Metaplectic Operators and FIOs

The $\hbar$-Weyl quantization $\sigma^{w,\hbar}$ of a generalized symbol $\sigma \in \mathcal{S}'(\mathbb{R}^{2d})$ is defined via duality by

$$\langle \sigma^{w,\hbar} f, g \rangle := \langle \sigma, W^{\hbar}(g, f) \rangle, \quad f, g \in \mathcal{S}(\mathbb{R}^d).$$

An explicit representation can be formally derived:

$$\sigma^{w,\hbar} f(x) = (2\pi\hbar)^{-d} \int_{\mathbb{R}^{2d}} e^{\frac{i}{\hbar}(x-y)\cdot\xi} \sigma\left(\frac{x+y}{2}, \xi\right) f(y) \, dy d\xi$$

$$= \int_{\mathbb{R}^{2d}} e^{2\pi i (x-y)\cdot\xi} \sigma\left(\frac{x+y}{2}, \hbar\xi\right) f(y) \, dy d\xi.$$

It is basically a matter of computation to show that most of the properties satisfied by the standard Weyl quantization (e.g., the Moyal formula, the symplectic covariance, etc.) are also satisfied by the $\hbar$-Weyl quantization after slight modifications in order to keep track of $\hbar$. In fact, the results in the previous chapters mostly refer to the case where $2\pi\hbar = 1$ in this general setting; we refer to [61, 106] for further details.

We now deal with $\hbar$-dependent metaplectic operators. Any operator $\mu(S) \in \mathrm{Mp}(d, \mathbb{R})$ associates with its semiclassical counterpart

$$\mu^{\hbar}(S) := D_{h^{-1/2}} \mu(S) D_{h^{1/2}}.$$

Note that $\mu^{\hbar}(S) \in \mathrm{Mp}(d, \mathbb{R})$, since the dilation D_{λ} is a metaplectic operator as recalled at the beginning of this section.

Let us focus on the class of quadratic Fourier transforms, that is metaplectic operators associated with free symplectic matrices. In view of the results stated in

Theorem 4.3.7, if $S = \begin{bmatrix} A & B \\ C & D \end{bmatrix} \in \mathrm{Sp}_0(d, \mathbb{R})$ (namely $\det B \neq 0$), then

$$\mu^{\hbar}(S) f(x) = c(2\pi\hbar)^{-d/2} |\det B|^{-1/2} \int_{\mathbb{R}^d} e^{\frac{i}{\hbar} \Phi_S(x,y)} f(y) dy, \qquad f \in S(\mathbb{R}^d),$$

(5.3)

for some $c \in \mathbb{C}$ with $|c| = 1$, where Φ_S is the generating function associated with S:

$$\Phi_S(x, y) := \frac{1}{2} DB^{-1} x \cdot x - B^{-1} x \cdot y + \frac{1}{2} B^{-1} Ay \cdot y.$$

Let us finally consider semiclassical Fourier integral operators. We refer to the notions and the notation introduced in Sects. 4.4 and 4.4.2 above.

Definition 5.2.1 Let χ be a tame canonical transformation and fix $s \geq 0$. We denote by $FIO_\hbar(\chi, v_s)$ the space of linear continuous operators $T : S(\mathbb{R}^d) \to S'(\mathbb{R}^d)$ such that $D_{\hbar^{1/2}} T D_{\hbar^{-1/2}} \in FIO(\chi, v_s)$. We also set

$$FIO_\hbar(\chi) := \bigcap_{s \geq 0} FIO_\hbar(\chi, v_s),$$

endowed with the seminorms

$$\|T\|_{m,\chi}^{\hbar} := \|D_{\hbar^{1/2}} T D_{\hbar^{-1/2}}\|_{m,\chi}, \qquad m \geq 0.$$

Similarly we denote by $FIO_\hbar'(S)$, $S \in \mathrm{Sp}(d, \mathbb{R})$, the space of linear continuous operators $T : S(\mathbb{R}^d) \to S'(\mathbb{R}^d)$ such that $D_{\hbar^{1/2}} T D_{\hbar^{-1/2}} \in FIO'(S)$.

From Theorems 4.4.2 and 4.4.3 we obtain at once the following results.

Theorem 5.2.2 Let $T \in FIO_\hbar(\chi)$. Then T extends to a bounded operator on $M_\hbar^p(\mathbb{R}^d)$, $1 \leq p \leq \infty$. Moreover, for every $m > 2d$ there exists a constant $C > 0$ depending only on m and the dimension d such that

$$\|T\|_{M_\hbar^p \to M_\hbar^p} \leq C \|T\|_{m,\chi}^{\hbar}.$$

Corollary 5.2.3 Let $T \in FIO_\hbar(\chi)$, $1 < p < \infty$ and $r = 2d|1/p - 1/2|$. Then T extends to a bounded operator $T : H_\hbar^{r,p}(\mathbb{R}^d) \to L^p(\mathbb{R}^d)$ if $1 < p \leq 2$ and $T : L^p(\mathbb{R}^d) \to H_\hbar^{-r,p}(\mathbb{R}^d)$ for $2 \leq p < \infty$. Moreover, for every $m > 2d$ there exists a constant $C > 0$ depending on m, d, p such that

$$\|T\|_{H_\hbar^{r,p} \to L^p} \leq C \|T\|_{m,\chi}^{\hbar} \qquad \text{for } 1 < p \leq 2$$

and

$$\|T\|_{L^p \to H_\hbar^{-r,p}} \le C \|T\|_{m,\chi}^{\hbar} \quad \text{for } 2 \le p < \infty.$$

Proof The results follow at once from Theorem 5.2.2 and Theorem 5.1.1. For example, for $1 < p \le 2$ we have

$$\|Tf\|_{L^p} \le C_1 h^{d(1/p-1/2)/2} \|Tf\|_{M^{p,\hbar}} \le C_1 C_2 h^{d(1/p-1/2)/2} \|T\|_{m,\chi}^{\hbar} \|f\|_{M^{p,\hbar}}$$

$$\le C_1 C_2 C_3 \|T\|_{m,\chi}^{\hbar} \|f\|_{H_\hbar^{r,p}}.$$

$\square$

Theorem 5.2.4 *If $T^{(i)} \in FIO_\hbar(\chi_i)$, $i = 1, 2$, then the composition $T^{(1)}T^{(2)}$ is in $FIO_\hbar(\chi_1 \circ \chi_2)$. Moreover for every $m > 2d$ there exists a constant $C > 0$ depending only on m, the dimension d, and upper bounds for the first derivatives of χ_1 such that*

$$\|T^{(1)}T^{(2)}\|_{m,\chi_1 \circ \chi_2}^{\hbar} \le C \|T^{(1)}\|_{m,\chi_1}^{\hbar} \|T^{(2)}\|_{m,\chi_2}^{\hbar}.$$

The definition of $FIO_\hbar(\chi)$ given above has been introduced in [179] and it is slightly more general than other definitions appearing in the literature, such as the one in [61]. In fact, the latter ultimately amounts to consider a subclass of $FIO_\hbar(\chi^h)$ (defined as above) obtained by means of a suitable h-dilation of the canonical transformation χ, that is by setting $\chi^h(z) := h^{-1/2}\chi(h^{1/2}z)$, $z \in \mathbb{R}^{2d}$. Nevertheless, the class $FIO_\hbar(\chi^h)$ has a prominent role in our investigations. Indeed, fix $T_0 > 0$ and consider the Cauchy problem for the Schrödinger equation

$$\begin{cases} i\hbar \partial_t \psi = H(t, \cdot)^{\text{w},\hbar} \psi \\ \psi(s, x) = f(x), \end{cases} \tag{5.4}$$

where $t \in [0, T_0]$, $s \in [0, T_0]$ is the initial time and the Hamiltonian is the $\hbar$-Weyl quantization of an observable $H(t, z)$, $z \in \mathbb{R}^{2d}$, which is continuous with respect to $(t, z) \in [0, T_0] \times \mathbb{R}^{2d}$ and smooth with respect to z, satisfying

$$|\partial_z^\alpha H(t, z)| \le C_\alpha, \quad \forall |\alpha| \ge 2, \ z \in \mathbb{R}^{2d}, \ t \in [0, T_0]. \tag{5.5}$$

We denote by $\chi(t, s) : \mathbb{R}^{2d} \to \mathbb{R}^{2d}$ the corresponding Hamiltonian flow, namely

$$(x(t, s, y, \eta), \xi(t, s, y, \eta)) = \chi(t, s)(y, \eta)$$

is the solution of the classical equations of motion

$$\dot{x} = \nabla_\xi H(t, x, \xi), \qquad \dot{\xi} = -\nabla_x H(t, x, \xi)$$

with initial condition at time $t = s$ given by $x(s, s, y, \eta) = y$, $\xi(s, s, y, \eta) = \eta$. It is easily verified that $\chi(t, s)$ is a tame canonical transformation, the estimates in (4.19) being satisfied with constants $C_{\alpha,\beta}$ independent of $s, t \in [0, T_0]$. The following result clarifies the relevance of the class $FIO_\hbar(\chi^h)$ [61, Proposition 3.7].

Proposition 5.2.5 *Let $U(t, s)$ be the Schrödinger propagator for the Cauchy problem (5.4). We have $U(t, s) \in FIO_\hbar(\chi^h(t, s))$, uniformly with respect to $\hbar \in (0, 1]$, $s, t \in [0, T_0]$. As a consequence, $U(t, s)$ is bounded on $M_\hbar^p(\mathbb{R}^d)$ for any $1 \leq p \leq \infty$, with operator norm uniformly bounded with respect to $\hbar \in (0, 1]$, $s, t \in [0, T_0]$.*

Proof Consider the propagator $\tilde{U}(t, s)$ satisfying

$$\frac{i}{2\pi} \partial_t \tilde{U}(t, s) = \frac{1}{h} H(t, h^{1/2} \cdot)^w \tilde{U}(t, s), \quad \tilde{U}(s, s) = I,$$

and compare it with $U(t, s)$, which instead satisfies

$$i\hbar \partial_t U(t, s) = H(t, \cdot)^{w,\hbar} U(t, s), \quad U(s, s) = I.$$

It is not difficult to prove that the two propagators are related by the formula

$$U(t, s) = D_{h^{-1/2}} \tilde{U}(t, s) D_{h^{1/2}}.$$

Indeed, using the fact that D_λ is a metaplectic operator (see the beginning of this chapter) and the symplectic covariance of the Weyl quantization, we have

$$\begin{aligned} i\hbar \partial_t U(t, s) &= i\hbar \partial_t D_{h^{-1/2}} \tilde{U}(t, s) D_{h^{1/2}} \\ &= D_{h^{-1/2}} H(t, h^{1/2} x, h^{1/2} \xi)^w \tilde{U}(t, s) D_{h^{1/2}} \\ &= H(t, x, h\xi)^w D_{h^{-1/2}} \tilde{U}(t, s) D_{h^{1/2}} \\ &= H(t, x, h\xi)^w U(t, s) \\ &= H(t, x, \xi)^{w,\hbar} U(t, s). \end{aligned}$$

Now, it was proved in [214, Corollary 7.4] that $\tilde{U}(t, s)$ satisfies the estimate

$$|\langle \tilde{U}(t, s)\pi(z)g, \pi(w)g \rangle| \leq \frac{C_n}{v_n(w - \chi^h(t, s)(z))}, \quad z, w \in \mathbb{R}^{2d}, \tag{5.6}$$

for every $n \in \mathbb{N}$ and $g \in \mathcal{S}(\mathbb{R}^d) \setminus \{0\}$, where

$$\chi^h(t, s)(z) = h^{-1/2} \chi(t, s)(h^{1/2} z) \tag{5.7}$$

is the flow corresponding to the Hamiltonian

$$h^{-1}H(t, h^{1/2}z). \tag{5.8}$$

The constant C_n in (5.6) can be chosen independent of $\hbar$, because by (5.5) the derivatives of order ≥ 2 of the Hamiltonian (5.8) are bounded uniformly with respect to $t \in [0, T_0]$, $0 < \hbar \leq 1$, and the assumptions in [214, Section 7] are therefore satisfied uniformly with respect to $\hbar$—the same therefore holds for the conclusion of [214, Corollary 7.4].

As a consequence, $U(t, s) = D_{h^{-1/2}} \tilde{U}(t, s) D_{h^{1/2}} \in FIO_\hbar(\chi^h(t, s))$ uniformly with respect to $\hbar \in (0, 1]$, $s, t \in [0, T_0]$ and the desired continuity result follows from Theorem 5.2.2.   □

We end this section with the special case of a quadratic Hamiltonian.

Theorem 5.2.6 *Let $Q(x, \xi)$ be a real-valued homogeneous polyonomial of degree 2. Then*

$$e^{-i\frac{t}{\hbar}Q^{w,\hbar}} = c(t)\mu^\hbar(S_t)$$

for some phase factor $c(t)$ ($|c(t)| = 1$), where S_t is the Hamiltonian flow associated with Q, as in Sect. 4.3.3.

Proof It follows from the proof of Proposition 5.2.5 and Theorem 4.3.9 that

$$e^{-i\frac{t}{\hbar}Q^{w,\hbar}} = D_{h^{-1/2}}e^{-2\pi i t Q^w}D_{h^{1/2}} = c(t)D_{h^{-1/2}}\mu(S_t)D_{h^{1/2}} = c(t)\mu^\hbar(S_t),$$

which concludes the proof.   □

Part II
Analysis of Feynman Path Integrals

Chapter 6
Pointwise Convergence of the Integral Kernels

6.1 Summary

In this chapter we address the problem of the pointwise convergence of the integral kernels of the Feynman-Trotter parametrices for the Schrödinger equation with a quadratic Hamiltonian perturbed by a pseudodifferential potential in suitable low regularity classes.

Precisely, consider the Cauchy problem for the Schrödinger equation

$$\begin{cases} i\hbar \partial_t \psi = H_0 \psi \\ \psi(0, x) = f(x), \end{cases}$$

where $H_0 = Q^{\mathrm{w}, \hbar}$ is the $\hbar$-Weyl quantization a real-valued, time-independent, quadratic homogeneous polynomial Q on $\mathbb{R}^{2d}$, namely

$$Q(x, \xi) = \frac{1}{2} A\xi \cdot \xi + Bx \cdot \xi + \frac{1}{2} Cx \cdot x,$$

for some $A, B, C \in \mathbb{R}^{d \times d}$ with $A = A^\top$ and $C = C^\top$. Note that a linear magnetic potential (uniform magnetic field) and a quadratic electric potential are possibly allowed and included in H_0.

It is proved in [142] that $H_0 = Q^{\mathrm{w}, \hbar}$ is a self-adjoint operator on the maximal domain

$$D(H_0) = \{f \in L^2(\mathbb{R}^d) : H_0 f \in L^2(\mathbb{R}^d)\}.$$

F. Nicola, S. I. Trapasso, *Wave Packet Analysis of Feynman Path Integrals*, Lecture Notes in Mathematics 2305, https://doi.org/10.1007/978-3-031-06186-8_6

We know from Theorem 5.2.6 that the associated propagator is a metaplectic operator, that is

$$U_0(t) = e^{-i\frac{t}{\hbar}H_0} = c(t)\mu^{\hbar}(S_t), \quad c(t) \in \mathbb{C}, \ |c(t)| = 1,$$

where the mapping

$$\mathbb{R} \ni t \mapsto S_t = \begin{bmatrix} A_t & B_t \\ C_t & D_t \end{bmatrix} \in \mathrm{Sp}(d, \mathbb{R})$$

is the phase-space flow determined by the Hamilton equations for the corresponding classical model with Hamiltonian $Q(x, \xi)$; we refer to Sect. 4.3.3 for an extensive account.

As observed in (5.3), if S_t is a free symplectic matrix, namely the upper-right block B_t of S_t is invertible, then the corresponding metaplectic operator coincides (up to a phase factor) with a quadratic Fourier transform. Hence

$$U_0(t)f(x) = c(t)(2\pi\hbar)^{-d/2}|\det B_t|^{-1/2} \int_{\mathbb{R}^d} e^{\frac{i}{\hbar}\Phi_t(x,y)} f(y)dy, \tag{6.1}$$

$(c(t) \in \mathbb{C}, |c(t)| = 1)$ where we introduced the quadratic form (also known as generating function of S_t, cf. (4.4))

$$\Phi_t(x, y) = \frac{1}{2}D_t B_t^{-1}x \cdot x - B_t^{-1}x \cdot y + \frac{1}{2}B_t^{-1}A_t y \cdot y. \tag{6.2}$$

For future reference we define the set of *exceptional times* as

$$\mathfrak{E} = \{t \in \mathbb{R} \ : \ \det B_t = 0\}, \tag{6.3}$$

namely the values of t such that S_t is *not* a free symplectic matrix. Some of the properties of this set are straightforward implications of the fact that it coincides with the zero set of an analytic function: aside from the case $\mathfrak{E} = \mathbb{R}$ (which trivially happens when $H_0 = 0$), $\mathfrak{E}$ is a discrete (hence at most countable) subset of $\mathbb{R}$ which always includes $t = 0$—and $\mathfrak{E} = \{0\}$ in the case of the free Schrödinger equation.

We now consider the perturbed problem

$$\begin{cases} i\hbar\partial_t \psi = (H_0 + V)\psi \\ \psi(0, x) = f(x), \end{cases} \tag{6.4}$$

where we included the potential perturbation $V \in \mathcal{L}(L^2(\mathbb{R}^d))$. We are in the position to use the Trotter product formula in the form of Theorem 6.4.1 below: if $U(t) = e^{-i\frac{t}{\hbar}(H_0+V)}$ denotes the evolution operator associated with (6.4), then

$$U(t)f = \lim_{n \to \infty} E_n(t)f, \quad \forall f \in L^2(\mathbb{R}^d),$$

where the Feynman-Trotter approximate propagators $E_n(t)$ are defined by

$$E_n(t) := \left(e^{-\frac{i}{\hbar}\frac{t}{n}H_0}e^{-\frac{i}{\hbar}\frac{t}{n}V}\right)^n, \quad n \in \mathbb{N}, \, n \geq 1. \tag{6.5}$$

We will denote by $e_{n,t}(\hbar, x, y)$ the distribution kernel of $E_n(t)$ and by $u_t(\hbar, x, y)$ that of $U(t)$.

Now, we study the problem of the convergence of $e_{n,t}(\hbar, x, y)$ to $u_t(\hbar, x, y)$ as $n \to +\infty$. To this end, we specialize the potential V in suitable classes. Consider first the case when V is just the pointwise multiplication by a function $V(x)$. We consider then the following spaces.

1. The best option for our purposes is given by the Hörmander class $C_b^\infty(\mathbb{R}^d)$, the space of smooth bounded functions on $\mathbb{R}^d$ with bounded derivatives of any order.
2. At an intermediate level we have the (scale of) modulation spaces $M_{0,s}^\infty(\mathbb{R}^d)$, $s > 2d$, consisting (see Definition 3.2.1) of distributions $f \in S'(\mathbb{R}^d)$ such that for any $g \in S(\mathbb{R}^d) \setminus \{0\}$

$$|V_g f(x, \xi)| \leq C(1 + |\xi|)^{-s}, \quad (x, \xi) \in \mathbb{R}^{2d},$$

   for some $C > 0$. $M_{0,s}^\infty(\mathbb{R}^d)$ contain bounded continuous functions, which become less and less regular as $s \searrow 2d$—the parameter s can be rightfully thought of as a degree of (fractional) differentiability.
3. We finally consider the Sjöstrand class $M^{\infty,1}(\mathbb{R}^d)$ as a maximal space, where the partial regularity of the previous level is completely lost. Recall from Sect. 3.5 that $f \in M^{\infty,1}(\mathbb{R}^d)$ if for any $g \in S(\mathbb{R}^d) \setminus \{0\}$

$$\|f\|_{M^{\infty,1}} = \int_{\mathbb{R}^d} \sup_{x \in \mathbb{R}^d} |V_g f(x, \xi)| d\xi < \infty.$$

It is still a space of bounded continuous functions which locally enjoy the mild regularity of the Fourier transform of a L^1 function.

We have indeed the following chain of strict inclusions for $s > d$ (cf. Proposition 3.5.1):

$$C_b^\infty(\mathbb{R}^d) \subset M_{0,s}^\infty(\mathbb{R}^d) \subset M^{\infty,1}(\mathbb{R}^d) \subset (\mathcal{F}L^1(\mathbb{R}^d))_{\text{loc}} \cap L^\infty(\mathbb{R}^d) \subset C_b(\mathbb{R}^d).$$

It seems worthwhile to highlight that results on the convergence of path integrals are already known for special elements of the Sjöstrand class. For instance, a family of potentials extensively investigated from different angles in the work of S. Albeverio and co-authors [1–4] and K. Itô [146, 147] is given by the space $\mathcal{F}M(\mathbb{R}^d)$ of Fourier transforms of (finite) complex measures on $\mathbb{R}^d$. In fact, we have $\mathcal{F}M(\mathbb{R}^d) \subset M^{\infty,1}(\mathbb{R}^d)$, cf. Proposition 3.5.1, and the above inclusion is strict; for instance, $f(x) = \cos|x|$, $x \in \mathbb{R}^d$, clearly belongs to $C_b^\infty(\mathbb{R}^d)$, but it is easy to realize that $f \notin \mathcal{F}M(\mathbb{R}^d)$ as soon as $d > 1$, by the known formula for the fundamental solution of the wave equation [79].

In fact, we will cover a more general family of potential perturbations in the form of Weyl operators, namely we assume that $V = \sigma^{w,\hbar}$, where the symbol σ belongs to any of the spaces $C_b^\infty(\mathbb{R}^{2d})$, $M_{0,s}^\infty(\mathbb{R}^{2d})$ with $s > 2d$, $M^{\infty,1}(\mathbb{R}^{2d})$. Notice indeed that the multiplication by a function V on $\mathbb{R}^d$ is an easy example of Weyl operator with symbol $\sigma_V(x,\xi) = V(x) = (V \otimes 1)(x,\xi)$, and the correspondence $V \mapsto \sigma$ is continuous from $M^{\infty,1}(\mathbb{R}^d)$ to $M^{\infty,1}(\mathbb{R}^{2d})$ and similarly for the other spaces mentioned above, cf. Remark 4.2.1. We will thus exploit the very rich structure enjoyed by the modulation spaces $M_{0,s}^\infty(\mathbb{R}^{2d})$ (with $s > 2d$) and $M^{\infty,1}(\mathbb{R}^{2d})$: recall that they are Banach algebras for both pointwise multiplication and Weyl product of symbols (cf. Remark 4.2.3), and the corresponding families of Weyl operators are inverse-closed Banach subalgebras of $\mathcal{L}(L^2(\mathbb{R}^d))$ (cf. Theorem 4.2.4).

After a brief collection of preliminary results (Sects. 6.2 and 6.3), in Sect. 6.4 we provide a representation the propagator $U(t)$ associated with the problem (6.4) as a member of suitable FIO classes introduced in Sect. 4.4 above; moreover, we give a direct proof of the Trotter product formula in the setting of the present chapter. The subsequent Sects. 6.5, 6.6, and 6.7 are devoted to the claimed pointwise convergence result at the level of integral kernels for the above Feynman-Trotter parametrices $E_n(t)$, with potentials at the different regularity levels encoded by symbols in the spaces $M_{0,s}^\infty(\mathbb{R}^{2d})$ with $s > 2d$, $C_b^\infty(\mathbb{R}^{2d})$ and $M^{\infty,1}(\mathbb{R}^{2d})$ respectively. The occurrence of exceptional times is a problem that we addressed in Sect. 6.8, where weaker convergence results are provided, while the concluding Sect. 6.9 offers some hints on a possible physical interpretation of this phenomenon.

6.2 Preliminary Results

6.2.1 The Schwartz Kernel Theorem

The Schwartz kernel theorem is a cornerstone of the theory of generalized functions, usually summoned in order to prove that a fairly well-behaved operator has an explicit representation as an integral transform—in the sense of distributions. We will resort below to this crucial identification, but first we need to carefully specify the underlying topological details [115, 223]. Recall from Sect. 2.2 that $\mathcal{S}'(\mathbb{R}^d)$ and $\mathcal{S}'(\mathbb{R}^{2d})$ are endowed with the strong topology by default.

A linear map $A: S(\mathbb{R}^d) \to S'(\mathbb{R}^d)$ is continuous if and only if it is generated by a (unique) temperate distribution $k \in S'(\mathbb{R}^{2d})$, namely:

$$\langle Af, g \rangle = \langle k, g \otimes \overline{f} \rangle, \qquad \forall f, g \in S(\mathbb{R}^d),$$

and the correspondence $k \mapsto A$ above is a topological isomorphism between $S'(\mathbb{R}^{2d})$ and the space $\mathcal{L}_b\left(S(\mathbb{R}^d), S'(\mathbb{R}^d)\right)$.

The previous identification provides the following convergence result for the distribution kernels.

Proposition 6.2.1 *Let $A_n \to A$ in $\mathcal{L}_s\left(S(\mathbb{R}^d), S'(\mathbb{R}^d)\right)$. Then we have convergence in $S'(\mathbb{R}^{2d})$ of the corresponding distribution kernels.*

Proof Since $S(\mathbb{R}^d)$ is a Fréchet space and A_n, being a sequence, defines a filter with countable basis on $\mathcal{L}_s(S(\mathbb{R}^d), S'(\mathbb{R}^d))$, from the Banach-Steinhaus theorem [223, Corollary at pag. 348] we have that $A_n \to A$ also in $\mathcal{L}_c(S(\mathbb{R}^d), S'(\mathbb{R}^d))$, which is in turn equivalent to convergence in $\mathcal{L}_b(S(\mathbb{R}^d), S'(\mathbb{R}^d))$ since $S(\mathbb{R}^d)$ is a Montel space—cf. [223, Propositions 34.4 and 34.5]. The desired conclusion then follows from the Schwartz kernel theorem. $\qquad\qquad\qquad\qquad\qquad\qquad\qquad\qquad \square$

6.2.2 Uniform Estimates for Linear Changes of Variable

The following technical lemma extends a result in [55, Lemma 2.2].

Lemma 6.2.2 *Let X denote either $M^{\infty}_{0,s}(\mathbb{R}^{2d})$, $s \geq 0$, or $M^{\infty,1}(\mathbb{R}^{2d})$. Let $\sigma \in X$ and $t \mapsto S_t \in \mathrm{Sp}\,(d, \mathbb{R})$ be a continuous mapping defined on the compact interval $[-T, T]$, $T > 0$. For any $t \in [-T, T]$, we have $\sigma \circ S_t \in X$, with*

$$\|\sigma \circ S_t\|_X \leq C(T)\,\|\sigma\|_X.$$

Proof The case $X = M^{\infty,1}(\mathbb{R}^{2d})$ is covered by [55, Lemma 2.2]. We prove here the claim for $X = M^{\infty}_{0,s}(\mathbb{R}^{2d})$.

For any non-zero window function $\Phi \in S\left(\mathbb{R}^{2d}\right)$ and $S \in \mathrm{Sp}\,(d, \mathbb{R})$ we have

$$
\begin{aligned}
\|\sigma \circ S\|_{M^{\infty}_{0,s}} &= \sup_{z,\zeta \in \mathbb{R}^{2d}} \left| \langle \sigma \circ S, M_\zeta T_z \Phi \rangle \right| v_s(\zeta) \\
&= \sup_{z,\zeta \in \mathbb{R}^{2d}} \left| \left\langle \sigma, M_{(S^{-1})^\top \zeta} T_{Sz}\left(\Phi \circ S^{-1}\right) \right\rangle \right| v_s(\zeta) \\
&= \sup_{z,\zeta \in \mathbb{R}^{2d}} \left| \left\langle \sigma, M_\zeta T_z\left(\Phi \circ S^{-1}\right) \right\rangle \right| v_s(S^\top \zeta) \\
&\leq \left\| S^\top \right\|^s \left\| V_{\Phi \circ S^{-1}} \sigma \right\|_{M^{\infty}_{0,s}} \\
&\lesssim \|S\|^s \left\| V_{\Phi \circ S^{-1}} \Phi \right\|_{L^1_s} \|\sigma\|_{M^{\infty}_{0,s}},
\end{aligned}
$$

where we used the estimate $v_s(S^\top \zeta) \leq \|S^\top\|^s v_s(\zeta)$ (here $\|\mathcal{B}\|$ denotes the operator norm of the matrix $\mathcal{B}$) and the change-of-window formula (3.1.2).

We now prove the uniformity with respect to the parameter t, when $S = S_t$. The subset $\{S_t : t \in [-T, T]\} \subset \mathrm{Sp}(d, \mathbb{R})$ is bounded and thus $\|S_t\| \leq C_1(T)$. Furthermore, $\{V_{\Phi \circ S_t^{-1}} \Phi : t \in [-T, T]\}$ is a bounded subset of $\mathcal{S}(\mathbb{R}^{2d})$ (this follows at once by inspecting the Schwartz seminorms of $\Phi \circ S_t^{-1}$), hence $\|V_{\Phi \circ S^{-1}} \Phi\|_{L_s^1} \leq C_2(T)$.   □

6.2.3 Exponentiation in Banach Algebras

We prove now an easy result on exponentials in Banach algebras. Recall the notation introduced in Remark 2.5.1.

Lemma 6.2.3 *Let* $(A, \star)$ *be a complex Banach algebra with unit* $\mathbb{1}$ *and consider* $a \in A$. *For any real* t *and integer* $n \geq 1$ *we have*

$$e^{-i\frac{t}{n}a} := \sum_{k=0}^{\infty} \left(-i\frac{t}{n}\right)^k \frac{a^k}{k!} = \mathbb{1} + i\frac{t}{n}a_0,$$

where $a_0 \in A$ *and the following estimate holds:*

$$\|a_0\| \leq \|a\|\, e^{|t|\|a\|}.$$

Proof It is enough to set

$$a_0 := -\sum_{k=0}^{\infty} \left(-i\frac{t}{n}\right)^k \frac{a^{k+1}}{(k+1)!}.$$

The desired identity is clearly satisfied and we can estimate the norm of a_0 as follows:

$$\|a_0\| \leq \|a\| \left(\sum_{k=0}^{\infty} \frac{|t|^k \|a\|^k}{(k+1)!}\right)$$

$$= \frac{1}{|t|}\left(e^{|t|\|a\|} - 1\right) \leq \|a\|\, e^{|t|\|a\|}.$$

□

We will repeatedly make use of the following result, which is an easy consequence of Theorem 4.2.4 and the previous lemma.

Corollary 6.2.4 *Let* $X = M_{0,s}^{\infty}(\mathbb{R}^{2d})$, $s > 2d$, *or* $X = M^{\infty,1}(\mathbb{R}^{2d})$, *so that* $(X, \#)$ *is a Banach algebra under the Weyl product and* $\{\mathrm{op_w}(\sigma) : \sigma \in X\}$ *is a subalgebra of* $\mathcal{L}(L^2(\mathbb{R}^d))$ *under composition. The Weyl quantization* $\mathrm{op_w} \colon X \to \mathcal{L}(L^2(\mathbb{R}^d))$ *is a homomorphism of Banach algebras. In particular, for any* $\sigma \in X$, $t \in \mathbb{R}$ *and* $n \in \mathbb{N}$ *we have*

$$e^{-i\frac{t}{n}\sigma^{\mathrm{w}}} = (e^{-i\frac{t}{n}\sigma})^{\mathrm{w}} = I + i\frac{t}{n}\sigma_0^{\mathrm{w}},$$

where $\sigma_0 \in X$ *satisfies*

$$\|\sigma_0\|_X \leq \|\sigma\|_X e^{|t|\|\sigma\|_X}.$$

6.2.4 Two Technical Lemmas

We collect here two results that will be used in the proof of Theorem 6.7.1 below.

The first one is a factorization result in Banach algebras that can be easily proved by induction on n.

Lemma 6.2.5 *Let* A *be a Banach algebra. For any* $u_1, \ldots, u_n, v_1, \ldots, v_n \in A$, *with* $\|u_i\| \leq R$ *and* $\|v_i\| \leq S$ *for any* $i = 1, \ldots, n$ *and some* $R, S > 0$, *and setting* $w_k = u_k + v_k$, *we have*

$$\prod_{k=1}^{n} (u_k + v_k) = u_1 u_2 \ldots u_n + z_n,$$

where

$$z_n = v_1 w_2 \ldots w_n + u_1 v_2 w_3 \ldots w_n + \ldots + u_1 u_2 \ldots u_{n-2} v_{n-1} w_n + u_1 u_2 \ldots u_{n-1} v_n,$$

and therefore

$$\|z_n\| \leq nS(R+S)^{n-1}.$$

Lemma 6.2.6 *Let* F_n *and* G_n *be two sequences of complex-valued functions on* $\mathbb{R}^{2d}$ *such that* $F_n \to F$, $G_n \to G$ *pointwise, and assume* $|F_n| \leq H \in L^1(\mathbb{R}^{2d})$ *and* $\|G_n\|_{L^1} \leq \epsilon$ *for any* $n \in \mathbb{N}$. *Then,*

$$\limsup_{n \to \infty} \|F_n + G_n - (F + G)\|_{L^1} \leq 2\epsilon.$$

Proof First, notice that $\|G\|_{L^1} \leq \epsilon$ by Fatou's lemma. Then

$$\|F_n + G_n - (F + G)\|_{L^1} \leq \|F_n - F\|_{L^1} + \|G_n - G\|_{L^1},$$

where the first term on the right-hand side vanishes by dominated convergence, while the remaining one satisfies $\|G_n - G\|_{L^1} \leq 2\epsilon$. The claim is thus proved. $\quad\square$

6.3   Reduction to the Case $\hbar = (2\pi)^{-1}$

We are dealing with the perturbed problem (6.4), where $H_0 = Q^{w,\hbar}$ is the $\hbar$-Weyl quantization of a real-valued quadratic form on $\mathbb{R}^{2d}$ and $V = \sigma^{w,\hbar}$ is a $\hbar$-Weyl operator with symbol σ in the classes $M_{0,s}^{\infty}(\mathbb{R}^{2d})$, $s > 2d$, or $C_b^{\infty}(\mathbb{R}^{2d})$ or $M^{\infty,1}(\mathbb{R}^{2d})$. As observed in the summary, we are interesting to the convergence of the Feynman-Trotter propagators $E_n(t)$ defined in (6.5).

Although we will state the results of this chapter for a general value of $\hbar > 0$, for simplicity we will carry out the proofs for $\hbar = (2\pi)^{-1}$. In fact, one can reduce to that case by means of a simple argument that we are going to anticipate here, since it applies to all the subsequent results.

Firstly, we have the following formulas, that are a consequence of the discussion in the proof of Proposition 5.2.5:

$$\left(e^{-\frac{i}{\hbar}\frac{t}{n}Q^{w,\hbar}} e^{-\frac{i}{\hbar}\frac{t}{n}\sigma^{w,\hbar}}\right)^n = D_{h^{-1/2}}\left(e^{-2\pi i\frac{t}{n}Q^w} e^{-2\pi i\frac{t}{n}\tilde{\sigma}^w}\right)^n D_{h^{1/2}}$$

where $\tilde{\sigma}(x,\xi) = h^{-1}\sigma(h^{1/2}x, h^{1/2}\xi)$. Similarly,

$$e^{-\frac{i}{\hbar}t(Q^{w,\hbar}+\sigma^{w,\hbar})} = D_{h^{-1/2}}e^{-2\pi it(Q^w+\tilde{\sigma}^w)}D_{h^{1/2}}.$$

Secondly, all the function spaces involved in the results in this chapter are invariant with respect to dilations (in particular, if σ belongs to $M_{0,s}^{\infty}(\mathbb{R}^{2d})$, $s > 2d$, or $C_b^{\infty}(\mathbb{R}^{2d})$ or $M^{\infty,1}(\mathbb{R}^{2d})$, the same holds for $\tilde{\sigma}$) and no uniformity issue is concerned because $\hbar > 0$ will be always arbitrary but fixed.

Therefore, *in the proofs* of the results in this chapter (except in Sects. 6.8 and 6.9, where there would not be any particular advantage) we will set $\hbar = (2\pi)^{-1}$, hence

$$H_0 = Q^{w,\hbar} = Q^w, \quad V = \sigma^{w,\hbar} = \sigma^w.$$

We simply denote by $e_{n,t}(x,y)$ the distribution kernel of

$$E_n(t) = \left(e^{-2\pi i\frac{t}{n}Q^w} e^{-2\pi i\frac{t}{n}\sigma^w}\right)^n = \left(e^{-2\pi i\frac{t}{n}H_0} e^{-2\pi i\frac{t}{n}V}\right)^n$$

and by $u_t(x, y)$ that of

$$U(t) = e^{-2\pi it(Q^{\mathrm{w}} + \sigma^{\mathrm{w}})} = e^{-2\pi it(H_0 + V)}.$$

6.4   The Fundamental Solution and the Trotter Formula

Here we address the problem of finding an integral representation of the propagator for the perturbed problem (6.4), namely we construct its fundamental solution, for potentials in the classes considered in the summary. The relevant result was proved in [55] and gives a representation of the propagator as an operator in the classes of Fourier integral operators introduced in Sect. 4.4.2. It represents a far-reaching generalization of previous work [233], where special classes of *smooth* symbols were considered. We provide a proof along the same lines as that in [55], but conceptually simpler. We also provide a proof of well-posedness in L^2 and the Trotter formula for the same problem for a general $V \in \mathcal{L}(L^2(\mathbb{R}^d))$.

Theorem 6.4.1   *Consider the problem* (6.4) *with* $H_0 = Q^{\mathrm{w},\hbar}$, *for a real-valued quadratic form* $Q(x, \xi)$, *and* $V \in \mathcal{L}(L^2(\mathbb{R}^d))$.

  (i) *The problem* (6.4) *is globally backward and forward well-posed in* $L^2(\mathbb{R}^d)$, *i.e. for every* $f \in L^2(\mathbb{R}^d)$ *there exists a unique solution in* $C(\mathbb{R}; L^2(\mathbb{R}^d))$ *and the propagator* $U(t) = e^{-\frac{i}{\hbar}(H_0 + V)}$ *defined by* $U(t)f := \psi(t, \cdot)$, $f \in L^2(\mathbb{R}^d)$, $t \in \mathbb{R}$, *is a one-parameter strongly continuous group of automorphisms of* $L^2(\mathbb{R}^d)$. *If* V *is self-adjoint,* $U(t)$ *is unitary.*
 (ii) *For every* $t \in \mathbb{R}$ *and* $\hbar > 0$,

$$\lim_{n \to +\infty} \left(e^{-\frac{i}{\hbar}\frac{t}{n}H_0} e^{-\frac{i}{\hbar}\frac{t}{n}V} \right)^n f = e^{-\frac{i}{\hbar}t(H_0 + V)} f, \quad \forall f \in L^2(\mathbb{R}^d).$$

(iii) *Let now* $V = \sigma^{\mathrm{w},\hbar}$ *with* $\sigma \in M^\infty_{0,s}(\mathbb{R}^{2d})$ *and* $s > 2d$. *For any* $t \in \mathbb{R}$, $\hbar > 0$ *the propagator* $U(t)$ *belongs to the class* $FIO_\hbar(S_t, v_s)$ (cf. *Definition* 5.2.1), *where* S_t *is the phase-space flow associated with the Hamiltonian* Q. *Moreover, if* $t \in \mathbb{R} \setminus \mathfrak{E}$ (cf. (6.3)) *it has an integral representation of the type*

$$U(t)f(x) = \int_{\mathbb{R}^d} e^{\frac{i}{\hbar}\Phi_t(x,y)} a(\hbar, t, x, y) f(y)\, dy,$$

for some amplitude $a(\hbar, t, \cdot) \in M^\infty_{0,s}(\mathbb{R}^{2d})$, *with* Φ_t *as in* (6.2).
   The same holds if we replace $M^\infty_{0,s}(\mathbb{R}^{2d})$ *by* $M^{\infty,1}(\mathbb{R}^{2d})$ *and* $FIO_\hbar(S_t, v_s)$ *by* $FIO'_\hbar(S_t)$.

Proof Without loss of generality we can set $\hbar = (2\pi)^{-1}$ (cf. the previous section).

(i) We work in the so-called interaction representation and we rephrase the problem in integral form. Namely, suppose $\psi \in C(\mathbb{R}; L^2(\mathbb{R}^d))$ is a distribution solution of the problem (6.4) and set

$$\psi(t, \cdot) = e^{-2\pi i t H_0} \varphi(t, \cdot).$$

Observe from the equation that $\psi \in C^1(\mathbb{R}; \mathcal{S}'(\mathbb{R}^d))$, and therefore by Theorem 4.3.8 the same holds for $\varphi(t, \cdot) = e^{2\pi i t H_0} \psi(t, \cdot)$, which will satisfy the problem

$$\begin{cases} \frac{i}{2\pi} \partial_t \varphi = V_t \varphi \\ \varphi(0, x) = f(x), \end{cases} \qquad (t, x) \in \mathbb{R} \times \mathbb{R}^d, \qquad (6.6)$$

where

$$V_t := e^{2\pi i t H_0} V e^{-2\pi i t H_0}.$$

Note indeed that V_t, $t \in \mathbb{R}$, defines a strongly continuous family of equi-bounded operators on $L^2(\mathbb{R}^d)$ as $\|V_t\|_{L^2 \to L^2} = \|V\|_{L^2 \to L^2}$. Therefore $\varphi \in C^1(\mathbb{R}; L^2(\mathbb{R}^d))$ and satisfies the following integral equation:

$$\varphi(t, \cdot) = f - 2\pi i \int_0^t V_s \varphi(s, \cdot) \, ds =: A\varphi,$$

where the integral is understood in the strong operator topology of $L^2(\mathbb{R}^d)$, i.e. as the limit of a sequence of Riemann sums convergent in $\mathcal{L}_s(L^2(\mathbb{R}^d))$. Vice versa, notice that if $\varphi \in C(\mathbb{R}; L^2(\mathbb{R}^d))$ is a solution of this integral equation then $\varphi \in C^1(\mathbb{R}; L^2(\mathbb{R}^d))$ and φ satisfies the Cauchy problem (6.6), so that $\psi(t, \cdot) = e^{-2\pi i t H_0} \varphi(t, \cdot)$ is a distribution solution in $C(\mathbb{R}; L^2(\mathbb{R}^d))$ of (6.4).

We then need to prove existence and uniqueness of the solution $\varphi \in C(\mathbb{R}; L^2(\mathbb{R}^d))$ for the above integral equation. Setting $U'(t)f := \varphi(t, \cdot)$ we will see that $U'(t)$ is a one-parameter strongly continuous family of bounded operators on $L^2(\mathbb{R}^d)$—unitary if V is self-adjoint—so that the same holds for the propagator $U(t) = e^{2\pi i t H_0} U'(t)$ of the original problem. Indeed, the group property follows by the uniqueness of the solution, as usual, and implies that $U(t)$ and $U'(t)$ are invertible for every $t \in \mathbb{R}$.

Precisely, consider the forward problem, i.e. $t \geq 0$; a similar argument applies to the backward problem. For arbitrary $T > 0$, we apply the Banach contraction theorem to the above defined operator A in the space

$C([0, T]; L^2(\mathbb{R}^d))$, with the norm (equivalent to the usual one)

$$\sup_{t \in [0,T]} e^{-\lambda t} \|\varphi(t, \cdot)\|_{L^2},$$

where $\lambda > \|V\|_{L^2 \to L^2}$. We deduce that there exists a unique solution $\varphi \in C([0, T]; L^2(\mathbb{R}^d))$ given by the usual Picard iteration scheme, which gives $\varphi(t, \cdot) = U'(t)f$, where $U'(t)$ is given by the so-called Dyson formula

$$U'(t) := I + \sum_{k=1}^{+\infty} (-2\pi i)^k \int_0^t \int_0^{t_1} \int_0^{t_2} \cdots \int_0^{t_{k-1}} V_{t_1} \circ V_{t_2} \circ \ldots \circ V_{t_k} \, dt_k \ldots dt_2 dt_1,$$

(6.7)

where the iterated integrals are again understood in the strong operator topology and the series converges in the same topology. However, an applicaton of the Minkowski integral inequality shows that the above series actually converges in $\mathcal{L}_b(L^2(\mathbb{R}^d))$. Indeed, we have

$$\left\| \int_0^t \int_0^{t_1} \int_0^{t_2} \cdots \int_0^{t_{k-1}} V_{t_1} \circ V_{t_2} \circ \ldots \circ V_{t_k} \, dt_k \ldots dt_2 dt_1 \right\|_{L^2 \to L^2} \leq \frac{C^k t^k}{k!},$$

(6.8)

(cf. [191, X.69] or [77, Chapter 3, Theorem 1.10]). This implies the existence of $U'(t)$ with the desired properties. Notice that if V is self-adjoint, V_t is self-adjoint as well for every t, so that from (6.6) we see that $\varphi(t, \cdot) = U'(t)f \in C^1(\mathbb{R}; L^2(\mathbb{R}^d))$ satisfies

$$\frac{d}{dt} \|\psi(t, \cdot)\|_{L^2}^2 = 0.$$

Therefore $U'(t)$ is an isometry of $L^2(\mathbb{R}^d)$, hence is unitary.

(ii) We use the telescopic identity

$$\left(e^{-2\pi i \frac{t}{n} H_0} e^{-2\pi i \frac{t}{n} V} \right)^n - e^{-2\pi i t (H_0 + V)} = \sum_{k=0}^{n-1} \left(e^{-2\pi i \frac{t}{n} H_0} e^{-2\pi i \frac{t}{n} V} \right)^k$$

$$\times \left(e^{-2\pi i \frac{t}{n} H_0} e^{-2\pi i \frac{t}{n} V} - e^{-2\pi i \frac{t}{n}(H_0 + V)} \right) \left(e^{-2\pi i \frac{t}{n}(H_0 + V)} \right)^{n-1-k}.$$

Observe that $\|e^{-2\pi i t H_0}\|_{L^2 \to L^2} = 1$, whereas a Taylor expansion in the Banach algebra $\mathcal{L}(L^2(\mathbb{R}^d))$ implies the fixed-time estimate

$\|e^{-2\pi i \frac{t}{n} V}\|_{L^2 \to L^2} \leq e^{Ct/n}$, for some constant $C > 0$. Therefore, for $f \in L^2(\mathbb{R}^d)$,

$$\left\| \left(e^{-2\pi i \frac{t}{n} H_0} e^{-2\pi i \frac{t}{n} V} \right)^n f - e^{-2\pi i t (H_0 + V)} f \right\|_{L^2} \leq n e^{Ct}$$

$$\times \sup_{s \in [0,1]} \left\| \left(e^{-2\pi i \frac{t}{n} H_0} e^{-2\pi i \frac{t}{n} V} - e^{-2\pi i \frac{t}{n} (H_0 + V)} \right) e^{-2\pi i s (H_0 + V)} f \right\|_{L^2}.$$

Since the set $\{e^{-2\pi i s (H_0 + V)} : s \in [0, 1]\} \subset L^2(\mathbb{R}^d)$ is compact, it is therefore sufficient to prove that

$$n \left(e^{-2\pi i \frac{t}{n} H_0} e^{-2\pi i \frac{t}{n} V} - e^{-2\pi i \frac{t}{n} (H_0 + V)} \right) \to 0 \quad \text{in } \mathcal{L}_c(L^2(\mathbb{R}^d)),$$

that is uniformly on the compact subsets of $L^2(\mathbb{R}^d)$. In fact, this is equivalent to pointwise convergence, as a consequence of the uniform boundedness principle and the fact that the topologies of pointwise convergence and of uniform convergence on compact subsets coincide on equicontinuous subsets of $L^2(\mathbb{R}^d)$, see [223, Corollary at page 348]. Hence, we have to prove that

$$n \left(e^{-2\pi i \frac{t}{n} H_0} e^{-2\pi i \frac{t}{n} V} - e^{-2\pi i \frac{t}{n} (H_0 + V)} \right) f \to 0 \quad \text{in } L^2(\mathbb{R}^d), \ \forall f \in L^2(\mathbb{R}^d),$$

which is in turn equivalent to

$$n \left(U' \left(\frac{t}{n} \right) - e^{-2\pi i \frac{t}{n} V} \right) f \to 0 \quad \text{in } L^2(\mathbb{R}^d), \quad \forall f \in L^2(\mathbb{R}^d),$$

where U' was defined in (6.7). By (6.7) and a Taylor expansion of $e^{-2\pi i \frac{t}{n} V}$ in $\mathcal{L}(L^2(\mathbb{R}^d))$ we get

$$n \left(U' \left(\frac{t}{n} \right) - e^{-2\pi i \frac{t}{n} V} \right) f = n \left(f - 2\pi i \int_0^{t/n} V_s f \, ds \right) - n \left(f - 2\pi i \frac{t}{n} V f \right)$$

$$+ O \left(\frac{1}{n} \right)$$

$$= 2\pi i n \int_0^{t/n} (V f - V_s f) ds + O \left(\frac{1}{n} \right).$$

By the fundamental theorem of calculus for integrals of continuous functions with values in $L^2(\mathbb{R}^d)$ we see that

$$n \int_0^{t/n} (V f - V_s f) ds \to 0 \quad \text{as} \quad n \to +\infty,$$

and this concludes the proof.

(iii) We consider the case $\sigma \in M^{\infty,1}(\mathbb{R}^{2d})$ (the case $\sigma \in M^{\infty}_{0,s}(\mathbb{R}^{2d})$, $s > 2d$ is similar).

Since $U(t) = e^{-2\pi i t H_0} U'(t)$ and $e^{-2\pi i t H_0} = c(t)\mu(S_t)$, $|c(t)| = 1$, the desired result will follow from Theorem 4.4.6 if we prove that the propagator $U'(t)$ for the problem (6.6) is given by a pseudodifferential operator with Weyl symbol in $M^{\infty,1}(\mathbb{R}^{2d})$. We define

$$\sigma_t^w := V_t = e^{2\pi i t H_0} V e^{-2\pi i t H_0} = e^{2\pi i t H_0} \sigma^w e^{-2\pi i t H_0}.$$

Let us observe that by the covariance property of the Weyl calculus (4.16) we have $\sigma_t(z) = \sigma(S_{-t} z)$, $z \in \mathbb{R}^{2d}$, and moreover

$$\|\sigma_t^w\|_{L^2 \to L^2} = \|\sigma^w\|_{L^2 \to L^2} \lesssim \|\sigma\|_{M^{\infty,1}}, \tag{6.9}$$

as a consequence of Theorem 4.2.2 (recall $M^{2,2} = L^2$).

We now have to prove that $U'(t)$ belongs to the Banach algebra $\mathcal{A}$ of pseudodifferential operators with Weyl symbol in $M^{\infty,1}$—endowed with the norm $\||a^w\|| := \|a\|_{M^{\infty,1}}$; see Remark 4.2.3 and Theorem 4.2.4. This is a consequence of (6.9) and the Banach algebra property, because each iterated integral in (6.7) is a pseudodifferential operator in $\mathcal{A}$, and by Lemma 6.4.2 below one can apply Minkowski integral inequality in $\mathcal{A}$, similarly to (6.8). Hence the series in (6.7) converges in $\mathcal{A}$ and the limit operator coincides with $U'(t)$ because $\mathcal{A} \hookrightarrow \mathcal{L}_b(L^2)$ (continuous inclusions) by Theorem 4.2.2.

□

It remains to prove the following Lemma, which justifies the application of the Minkowski integral inequality in the above algebra $\mathcal{A}$. Actually, some care has to be taken because the map $t \mapsto \sigma_t$ is not continuous in $M^{\infty,1}$ in general.

Lemma 6.4.2 *Let $\sigma_t \in \mathcal{S}'(\mathbb{R}^d)$, $t \in [0, T]$, be a family of temperate distributions such that the map $[0, T] \ni t \mapsto \sigma_t^w$ is continuous in $\mathcal{L}_s(\mathcal{S}(\mathbb{R}^d), \mathcal{S}'(\mathbb{R}^d))$. Then*

(i) *The map $[0, T] \ni t \mapsto \sigma_t$ is continuous in $\mathcal{S}'(\mathbb{R}^{2d})$.*
(ii) *The linear functional $\sigma : \mathcal{S}(\mathbb{R}^d) \to \mathbb{C}$, defined by*

$$\langle \sigma, \phi \rangle = \int_0^T \langle \sigma_t, \phi \rangle \, dt, \qquad \phi \in \mathcal{S}(\mathbb{R}^d) \tag{6.10}$$

is continuous, i.e. $\sigma \in \mathcal{S}'(\mathbb{R}^{2d})$ and we have

$$\langle \sigma^w f, g \rangle = \int_0^T \langle \sigma_t^w f, g \rangle \, dt, \qquad f, g \in \mathcal{S}(\mathbb{R}^d). \tag{6.11}$$

(iii) We have

$$\|\sigma\|_{M^{\infty,1}} \leq \int_0^T \|\sigma_t\|_{M^{\infty,1}} \, dt.$$

Proof

(i) Let k_t be the kernels of σ_t^w. It follows from Proposition 6.2.1 that the map $[0, T] \ni t \mapsto k_t$ is continuous in $S'(\mathbb{R}^{2d})$, and therefore the same applies to the map of the corresponding Weyl symbols σ_t.

(ii) As a consequence of item (i) and the uniform boundedness principle [223] the functionals σ_t are equicontinuous, namely $|\langle \sigma_t, \phi \rangle| \leq C p_n(\phi)$ for some constant C and some seminorm p_n in $S(\mathbb{R}^{2d})$ independent of t, which gives the desired continuity of σ.

  Formula (6.11) follows from (6.10) with ϕ replaced by $W(g, f)$, because $\langle \sigma^w f, g \rangle = \langle \sigma, W(g, f) \rangle$.

(iii) The equicontinuity of σ_t and the fact that the time-frequency shifts are strongly continuous in S [121] imply that $V_\phi \sigma_t = \langle \sigma_t, \pi(u, v)\phi \rangle$, $\phi \in S(\mathbb{R}^{2d})$, is continuous as a function of $(t, u, v) \in [0, T] \times \mathbb{R}^{2d} \times \mathbb{R}^{2d}$. Hence the function $(t, v) \mapsto \sup_{t \in [0,T]} |V_\phi \sigma_t(u, v)|$ is lower semicontinuous and therefore (Borel) measurable. The desired estimate then follows from (6.10) with $\pi(u, v)\phi$ in place of ϕ and Fubini's theorem.

$\square$

6.5 Potentials in $M_{0,s}^\infty$

In this section we address the problem of the pointwise convergence at the level of integral kernels for Feynman-Trotter parametrices $E_n(t)$ in (6.5), at the intermediate regularity encoded by $M_{0,s}^\infty$, $s > 2d$.

We denote by $e_{n,t}(\hbar, x, y)$ the distribution kernel of $E_n(t)$ and by $u_t(\hbar, x, y)$ that of $U(t) = e^{-it(H_0+V)}$.

Theorem 6.5.1 *Consider the problem* (6.4) *with* $H_0 = Q^{w,\hbar}$ *as discussed in the summary and* $V = \sigma^{w,\hbar}$ *with* $\sigma \in M_{0,s}^\infty(\mathbb{R}^{2d})$ *and* $s > 2d$. *For any fixed* $t \in \mathbb{R} \setminus \mathfrak{E}$ *(cf.* (6.3)*),* $\hbar > 0$:

(i) *The distributions* $e^{-\frac{i}{\hbar}\Phi_t} e_{n,t}$, $n \geq 1$, *and* $e^{-\frac{i}{\hbar}\Phi_t} u_t$ *belong to a bounded subset of* $M_{0,s}^\infty(\mathbb{R}^{2d})$ *(cf.* (6.2)*);*

(ii) $e_{n,t} \to u_t$ *in* $(\mathcal{F}L_r^1(\mathbb{R}^{2d}))_{\mathrm{loc}}$ *for any* $0 < r < s - 2d$, *hence uniformly on compact subsets.*

Remark 6.5.2 The first part of the claim ensures that the pointwise convergence problem makes sense in this case—the amplitudes are indeed bounded continuous functions. The second part provides a precise characterization of the regularity level at which convergence occurs, ultimately implying the desired pointwise convergence result.

Proof of Theorem 6.5.1 According with Sect. 6.3, we can assume $\hbar = (2\pi)^{-1}$, hence $H_0 = Q^w$ and $V = \sigma^w$ with $\sigma \in M_{0,s}^\infty(\mathbb{R}^{2d})$, with $s > 2d$. We also suppose $t > 0$, since the case $t < 0$ is similar. Actually, the upper-right block of the matrix $S_{-t} = S_t^{-1}$ is $-B_t^\top$ (cf. Proposition 4.3.1), hence $\det B_t \neq 0$ if and only if $\det B_{-t} \neq 0$.

In view Corollary 6.2.4 we have

$$E_n(t) = \left(e^{-2\pi i \frac{t}{n} H_0} e^{-2\pi i \frac{t}{n} V}\right)^n = \left(e^{-2\pi i \frac{t}{n} H_0}\left(I + 2\pi i \frac{t}{n}\sigma_0^w\right)\right)^n,$$

for a suitable $\sigma_0 = (\sigma_0)_{n,t} \in M_{0,s}^\infty(\mathbb{R}^{2d})$ satisfying

$$\|\sigma_0\|_{M_{0,s}^\infty} \leq C(t) \tag{6.12}$$

for some constant $C(t) > 0$ independent of n.

We use the symplectic covariance of Weyl calculus, that is we apply (4.13) and (4.16) repeatedly, and the fact that $e^{isH_0}e^{-isH_0} = I$ for any $s \in \mathbb{R}$ so that the ordered product of operators in $E_n(t)$ can be expanded as

$$E_n(t) = \left[\prod_{k=1}^n \left(I + 2\pi i \frac{t}{n}\left(\sigma_0 \circ S_{-k\frac{t}{n}}\right)^w\right)\right] e^{-2\pi it H_0}$$

$$= a_{n,t}^w\, e^{-2\pi it H_0},$$

and for any t and $n \geq 1$,

$$\|a_{n,t}\|_{M_{0,s}^\infty} = \left\|\prod_{k=1}^n \left(1 + 2\pi i \frac{t}{n}\left(\sigma_0 \circ S_{-k\frac{t}{n}}\right)\right)\right\|_{M_{0,s}^\infty}$$

$$\leq \prod_{k=1}^n \left(1 + 2\pi \frac{t}{n}\left\|\sigma_0 \circ S_{-k\frac{t}{n}}\right\|_{M_{0,s}^\infty}\right),$$

where in the first product symbol we mean the Weyl product # of symbols—cf. Sect. 4.2 and Remark 2.5.1. By Lemma 6.2.2 applied with $T = t$ and (6.12), we then have

$$\|a_{n,t}\|_{M_{0,s}^\infty} \leq \left(1 + \frac{t}{n}C(t)\right)^n \leq e^{C(t)t}, \tag{6.13}$$

for some new locally bounded constant $C(t) > 0$ independent of n.

Since S_t is a free symplectic matrix precisely for $t \in \mathbb{R} \setminus \mathfrak{E}$, by (4.13), (4.9) and Lemma 4.4.5 we explicitly have

$$E_n(t)\psi(x) = a_{n,t}^w \, e^{-2\pi i t H_0}\psi(x)$$

$$= c(t)\,|\det B_t|^{-1/2} \int_{\mathbb{R}^d} e^{2\pi i \Phi_t(x,y)}\widetilde{a_{n,t}}(x,y)\,\psi(y)\,dy,$$

where Φ_t is given in (6.2) and $c(t) \in \mathbb{C}$ is such that $|c(t)| = 1$.

Therefore, we managed to write $E_n(t)$ as an integral operator with kernel

$$e_{n,t}(x,y) = c(t)\,|\det B_t|^{-1/2}\, e^{2\pi i \Phi_t(x,y)}\widetilde{a_{n,t}}(x,y),$$

Now, consider the integral kernel u_t of the propagator $U(t) = e^{-2\pi i t(H_0+V)}$ and define for consistency $\widetilde{a}_t \in \mathcal{S}'(\mathbb{R}^{2d})$ in such a way that

$$u_t(x,y) = c(t)\,|\det B_t|^{-1/2}\, e^{2\pi i \Phi_t(x,y)}\widetilde{a}_t(x,y).$$

Since we know from the Trotter formula in Theorem 6.4.1 that for any fixed t

$$\|E_n(t)f - U(t)f\|_{L^2} \to 0, \qquad \forall f \in L^2(\mathbb{R}^d),$$

we have $E_n(t) \to U(t)$ in $\mathcal{L}_s(\mathcal{S}(\mathbb{R}^d), \mathcal{S}'(\mathbb{R}^d))$, because $\mathcal{S}(\mathbb{R}^d) \hookrightarrow L^2(\mathbb{R}^d) \hookrightarrow \mathcal{S}'(\mathbb{R}^d)$. As a consequence of Proposition 6.2.1, we get $e_{n,t} \to u_t$ in $\mathcal{S}'(\mathbb{R}^d)$. This is equivalent to

$$\widetilde{a_{n,t}} \to \widetilde{a}_t \quad \text{in } \mathcal{S}'(\mathbb{R}^{2d}).$$

Therefore, for any non-zero $\Psi \in \mathcal{S}(\mathbb{R}^{2d})$ we have pointwise convergence of the corresponding short-time Fourier transforms: for any fixed $(z,\zeta) \in \mathbb{R}^{4d}$,

$$V_\Psi \widetilde{a_{n,t}}(z,\zeta) = \langle \widetilde{a_{n,t}}, M_\zeta T_z \Psi \rangle \to \langle \widetilde{a}_t, M_\zeta T_z \Psi \rangle = V_\Psi \widetilde{a}_t(z,\zeta). \tag{6.14}$$

By (6.13) and Lemma 4.4.5 we see that the sequence $\widetilde{a_{n,t}}$, for any fixed t, is bounded in $M_{0,s}^\infty(\mathbb{R}^{2d})$. Hence, there exists a constant $C = C(t)$ independent of n such that

$$\left|V_\Psi \widetilde{a_{n,t}}(z,\zeta)\right| \le C \langle \zeta \rangle^{-s}, \qquad \forall z,\zeta \in \mathbb{R}^{2d}. \tag{6.15}$$

Combining this estimate with (6.14) immediately yields $\widetilde{a}_t \in M_{0,s}^\infty(\mathbb{R}^{2d})$ as well, hence the first claim in the statement.

To prove the second item of the claim we argue as follows. Fix a non-zero window $\Psi \in C_c^\infty(\mathbb{R}^{2d})$ and set $\Theta \in C_c^\infty(\mathbb{R}^{2d})$ with $\Theta = 1$ on $\mathrm{supp}\Psi$; for any fixed $z \in \mathbb{R}^{2d}$ and $0 < r < s - 2d$, we have

$$\left\| \mathcal{F}\left[(e_{n,t} - u_t) \, \overline{T_z \Psi} \right] \right\|_{L_r^1} = |\det B_t|^{-1/2} \left\| \mathcal{F}\left[e^{2\pi i \Phi_t} \left(\widetilde{a_{n,t}} - \widetilde{a}_t \right) \overline{T_z \Psi} \right] \right\|_{L_r^1}$$

$$= |\det B_t|^{-1/2} \left\| \mathcal{F}\left[\left(T_z \Theta e^{2\pi i \Phi_t} \right) \left(\widetilde{a_{n,t}} - \widetilde{a}_t \right) \overline{T_z \Psi} \right] \right\|_{L_r^1}$$

$$= |\det B_t|^{-1/2} \left\| \mathcal{F}\left[T_z \Theta e^{2\pi i \Phi_t} \right] * \mathcal{F}\left[\left(\widetilde{a_{n,t}} - \widetilde{a}_t \right) \overline{T_z \Psi} \right] \right\|_{L_r^1}$$

$$\lesssim |\det B_t|^{-1/2} \left\| \mathcal{F}\left[T_z \Theta e^{2\pi i \Phi_t} \right] \right\|_{L_r^1} \left\| \mathcal{F}\left[\left(\widetilde{a_{n,t}} - \widetilde{a}_t \right) \overline{T_z \Psi} \right] \right\|_{L_r^1},$$

the convolution inequality in the last step being an easy consequence of Peetre's inequality (2.1).

Clearly, $T_z \Theta e^{2\pi i \Psi_t} \in C_c^\infty(\mathbb{R}^{2d})$, while

$$\left\| \mathcal{F}\left[\left(\widetilde{a_{n,t}} - \widetilde{a}_t \right) \overline{T_z \Psi} \right] \right\|_{L_r^1} \to 0$$

by dominated convergence, using (6.14) and (recall that $s - r > 2d$)

$$\left| \mathcal{F}\left[\left(\widetilde{a_{n,t}} - \widetilde{a}_t \right) \overline{T_z \Psi} \right] \right| \langle \zeta \rangle^r = \left| V_\Psi \left(\widetilde{a_{n,t}} - \widetilde{a}_t \right) (z, \zeta) \right| \langle \zeta \rangle^r$$

$$\leq C \langle \zeta \rangle^{r-s} \in L^1(\mathbb{R}^{2d}),$$

where in the latter inequality we used (6.15) and the fact that $\widetilde{a}_t \in M_{0,s}^\infty(\mathbb{R}^{2d})$. The proof of convergence in $(\mathcal{F} L_r^1(\mathbb{R}^{2d}))_{\mathrm{loc}}$ is then complete.

To conclude, we have that

$$\left\| (e_{n,t} - u_t) \, \overline{T_z \Psi} \right\|_{L^\infty} \leq \left\| V_\Psi \left(e_{n,t} - u_t \right) (z, \cdot) \right\|_{L^1} \to 0,$$

which allows us to obtain uniform convergence on compact subsets: for any compact $K \subset \mathbb{R}^{2d}$, choose $\Psi \in \mathcal{S}(\mathbb{R}^{2d})$, $\Psi = 1$ on K and $z = 0$.

$\square$

6.6 Potentials in C_b^∞

We expect to improve the convergence result of the previous section in the smooth scenario. Indeed, we can state the following result.

Corollary 6.6.1 *Consider the problem* (6.4) *with* $H_0 = Q^{w,\hbar}$ *as discussed in the summary and* $V = \sigma^{w,\hbar}$ *with* $\sigma \in C_b^\infty(\mathbb{R}^{2d})$. *For any fixed* $t \in \mathbb{R} \setminus \mathfrak{E}$ *(cf.* (6.3)*),* $\hbar > 0$:

(i) *The distributions* $e^{-\frac{i}{\hbar}\Phi_t}e_{n,t}$, $n \geq 1$, *and* $e^{-\frac{i}{\hbar}\Phi_t}u_t$ *belong to a bounded subset of* $C_b^\infty(\mathbb{R}^{2d})$ *(cf.* (6.2)*);*

(ii) $e_{n,t} \to u_t$ *in* $C^\infty(\mathbb{R}^{2d})$, *hence uniformly on compact subsets together with any derivatives.*

It is worthwhile to compare this result with the second part of the claim in Fujiwara's Theorem 1.3.1. In spite of the different assumptions and approximation schemes, we stress that this result is global in time—more details on exceptional times are given below.

The proof of Corollary 6.6.1 is immediate, in view of the characterizations $C_b^\infty(\mathbb{R}^{2d}) = \bigcap_{s \geq 0} M_{0,s}^\infty(\mathbb{R}^{2d})$ (cf. Proposition 3.5.1) and

$$C^\infty(\mathbb{R}^{2d}) = \bigcap_{r>0}(\mathcal{F}L_r^1(\mathbb{R}^{2d}))_{\text{loc}}. \tag{6.16}$$

The latter characterization is folklore, being a refinement of the standard decay-smoothness trade-off for the Fourier transform; we provide a proof for the sake of completeness.

Proof of (6.16) The inclusion $C^\infty \subset \bigcap_{r>0}(\mathcal{F}L_r^1)_{\text{loc}}$ is straightforward. Indeed, let $f \in C^\infty$ and consider arbitrary $r > 0$ and $\phi \in C_c^\infty$; then $f\phi \in C_c^\infty$ and

$$\|f\phi\|_{\mathcal{F}L_r^1} = \int_{\mathbb{R}^{2d}} |\mathcal{F}(f\phi)(\zeta)|v_r(\zeta)d\zeta \lesssim_N \int_{\mathbb{R}^{2d}} (1+|\zeta|)^{-N+r}d\zeta,$$

and the latter quantity is finite for $N \in \mathbb{N}$ large enough.

Conversely, assume $f \in (\mathcal{F}L_r^1)_{\text{loc}}$ for any $r > 0$ and let $\phi \in C_c^\infty$. For any $\alpha \in \mathbb{N}$, the distribution derivatives $\partial^\alpha(f\phi)$ are in fact continuous functions, being the (inverse) Fourier transform of functions in L^1; indeed,

$$\int_{\mathbb{R}^{2d}} |\mathcal{F}(f\phi)(\zeta)|(1+|\zeta|)^{|\alpha|}d\zeta < \infty$$

by the assumption with $r = |\alpha|$. $\qquad\square$

6.7 Potentials in the Sjöstrand Class $M^{\infty,1}$

In this section we provide a finer convergence result, in the spirit of the previous ones, but for potentials in the Sjöstrand class.

Theorem 6.7.1 *Consider the problem (6.4) with $H_0 = Q^{w,\hbar}$ as discussed in the summary and $V = \sigma^{w,\hbar}$ with $\sigma \in M^{\infty,1}(\mathbb{R}^{2d})$. For any fixed $t \in \mathbb{R} \setminus \mathfrak{E}$ (cf. (6.3)), $\hbar > 0$:*

(i) *The distributions $e^{-\frac{i}{\hbar}\Phi_t} e_{n,t}$, $n \geq 1$, and $e^{-\frac{i}{\hbar}\Phi_t} u_t$ belong to a bounded subset of $M^{\infty,1}(\mathbb{R}^{2d})$ (cf. (6.2));*

(ii) *$e_{n,t} \to u_t$ in $(\mathcal{F}L^1(\mathbb{R}^{2d}))_{\mathrm{loc}}$, hence uniformly on compact subsets.*

Proof We assume $\hbar = (2\pi)^{-1}$—cf. Sect. 6.3. Now $H_0 = Q^w$ and $V = \sigma^w$ with $\sigma \in M^{\infty,1}(\mathbb{R}^{2d})$. Therefore for an arbitrary $\epsilon > 0$, Proposition 3.5.5 allows us to write $\sigma = \sigma_1 + \sigma_2$, with $\sigma_1 \in C_b^\infty(\mathbb{R}^{2d})$ and $\sigma_2 \in M^{\infty,1}(\mathbb{R}^{2d})$ with $\|\sigma_2\|_{M^{\infty,1}} \leq \epsilon$ and clearly

$$\|\sigma_1\|_{M^{\infty,1}} \leq \|\sigma\|_{M^{\infty,1}} + \|\sigma_2\|_{M^{\infty,1}} \leq \|\sigma\|_{M^{\infty,1}} + \epsilon \leq 1 + \|\sigma\|_{M^{\infty,1}},$$

assuming, from now on, $\epsilon \leq 1$. Notice that

$$e^{-2\pi i \frac{t}{n}(\sigma_1^w + \sigma_2^w)} = I + \sum_{k=1}^\infty \frac{1}{k!} \left(-2\pi i \frac{t}{n}\right)^k (\sigma_1^w + \sigma_2^w)^k$$

$$= I + 2\pi i \frac{t}{n}(\sigma_1')^w + 2\pi i \frac{t}{n}(\sigma_2')^w,$$

where we set

$$(\sigma_1')^w = -\sum_{k=1}^\infty \frac{1}{k!} \left(-2\pi i \frac{t}{n}\right)^{k-1} (\sigma_1^w)^k,$$

$$(\sigma_2')^w = -\sum_{k=1}^\infty \frac{1}{k!} \left(-2\pi i \frac{t}{n}\right)^{k-1} ((\sigma_1^w + \sigma_2^w)^k - (\sigma_1^w)^k).$$

Now, fix once for all $s > 2d$. The norms of the symbols $\sigma_1' = \sigma_{1,n,t}'$ and $\sigma_2' = \sigma_{2,n,t}'$ can be estimated as follows for any $t > 0$ (cf. the proof of Lemma 6.2.3). We have

$$\|\sigma_1'\|_{M^{\infty,1}} \leq \|\sigma_1\|_{M^{\infty,1}} e^{2\pi t \|\sigma_1\|_{M^{\infty,1}}} \leq (1 + \|\sigma\|_{M^{\infty,1}}) e^{2\pi t (1 + \|\sigma\|_{M^{\infty,1}})} =: C_1(t), \tag{6.17}$$

$$\|\sigma_1'\|_{M_{0,s}^\infty} \leq \|\sigma_1\|_{M_{0,s}^\infty} e^{2\pi t \|\sigma_1\|_{M_{0,s}^\infty}} =: C_2(t, \epsilon). \tag{6.18}$$

Similarly, using the elementary inequality

$$(a+b)^k - a^k \leq kb(a+b)^{k-1}, \qquad a, b \geq 0, \ k \geq 1,$$

we obtain

$$\|\sigma_2'\|_{M^{\infty,1}} \leq \|\sigma_2\|_{M^{\infty,1}} e^{2\pi t(\|\sigma_1\|_{M^{\infty,1}}+\|\sigma_2\|_{M^{\infty,1}})} \leq \epsilon e^{2\pi t(2+\|V\|_{M^{\infty,1}})} =: \epsilon\, C_3(t). \tag{6.19}$$

Here $C_1(t)$ and $C_3(t)$ are independent of n and ϵ and $C_2(t,\epsilon)$ is independent of n. The approximate propagator $E_n(t)$ thus becomes

$$E_n(t) = \left(e^{-2\pi i \frac{t}{n} H_0} e^{-2\pi i \frac{t}{n}(\sigma_1^{\mathrm{w}}+\sigma_2^{\mathrm{w}})} \right)^n$$

$$= \left(e^{-2\pi i \frac{t}{n} H_0} \left(1 + 2\pi i \frac{t}{n}(\sigma_1')^{\mathrm{w}} + 2\pi i \frac{t}{n}(\sigma_2')^{\mathrm{w}} \right) \right)^n,$$

and similar arguments to those in the proof of Theorem 6.5.1 yield

$$E_n(t) = \left[\prod_{k=1}^{n} \left(I + 2\pi i \frac{t}{n} \left(\sigma_1' \circ S_{-k\frac{t}{n}} \right)^{\mathrm{w}} + 2\pi i \frac{t}{n} \left(\sigma_2' \circ S_{-k\frac{t}{n}} \right)^{\mathrm{w}} \right) \right] e^{-2\pi i t H_0}$$

$$= \left[a_{n,t}^{\mathrm{w}} + b_{n,t}^{\mathrm{w}} \right] e^{-2\pi i t H_0},$$

where we set

$$a_{n,t} = \prod_{k=1}^{n} \left(1 + 2\pi i \frac{t}{n} \left(\sigma_1' \circ S_{-k\frac{t}{n}} \right) \right),$$

and in the latter product we mean the Weyl product # of symbols.

The term $a_{n,t}^{\mathrm{w}}$ can be estimated as in the proof of Theorem 6.5.1; in particular, using (6.18), we get (cf. (6.13))

$$\|a_{n,t}\|_{M^{\infty}_{0,s}} \leq C(t,\epsilon). \tag{6.20}$$

Setting

$$u_k = 1 + 2\pi i \frac{t}{n} \left(\sigma_1' \circ S_{-k\frac{t}{n}} \right), \qquad v_k = 2\pi i \frac{t}{n} \left(\sigma_2' \circ S_{-k\frac{t}{n}} \right), \quad k=1,\dots,n,$$

and applying Lemma 6.2.2 with $T = t$, and (6.17) and (6.19), we get

$$\|u_k\|_{M^{\infty,1}} = \left\| 1 + 2\pi i \frac{t}{n} \left(\sigma_1' \circ S_{-k\frac{t}{n}} \right) \right\|_{M^{\infty,1}} \leq 1 + 2\pi \frac{t}{n} \|\sigma_1' \circ S_{-k\frac{t}{n}}\|_{M^{\infty,1}} \leq 1 + \frac{t}{n} C(t),$$

$$\|v_k\|_{M^{\infty,1}} = 2\pi \frac{t}{n} \|\sigma_2' \circ S_{-k\frac{t}{n}}\|_{M^{\infty,1}} \leq \frac{t}{n} C(t)\epsilon,$$

for some locally bounded constant $C(t) > 0$ independent of n and ϵ. Therefore, by Lemma 6.2.5,

$$\left\|b_{n,t}\right\|_{M^{\infty,1}} \leq n\frac{t}{n}C(t)\epsilon\left(1+2\frac{t}{n}C(t)\right)^{n-1} \leq \epsilon t C(t)e^{2tC(t)}. \tag{6.21}$$

Following the pathway of the proof of Theorem 6.5.1, we write $E_n(t)$ as an integral operator with kernel

$$\begin{aligned} e_{n,t}(x,y) &= c(t)\left|\det B_t\right|^{-1/2} e^{2\pi i \Phi_t(x,y)}\left(\widetilde{a_{n,t}} + \widetilde{b_{n,t}}\right)(x,y) \\ &= c(t)\left|\det B_t\right|^{-1/2} e^{2\pi i \Phi_t(x,y)}k_{n,t}(x,y), \end{aligned}$$

where $c(t) \in \mathbb{C}$, $|c(t)| = 1$ and $k_{n,t} = \widetilde{a_{n,t}} + \widetilde{b_{n,t}}$, and the Trotter formula in Theorem 6.4.1 combined with Proposition 6.2.1 imply that $k_{n,t} \to k_t$ in $\mathcal{S}'(\mathbb{R}^{2d})$, where the distribution k_t is conveniently introduced to rephrase the integral kernel u_t of the propagator $U(t) = e^{-2\pi i t(H_0+V)}$ as

$$u_t(x,y) = c(t)\left|\det B_t\right|^{-1/2} e^{2\pi i \Phi_t(x,y)}k_t(x,y).$$

Observe, incidentally, that $k_t \in M^{\infty,1}(\mathbb{R}^{2d})$ by Theorem 6.4.1. By repeating the same argument with $\sigma_2 = 0$ (hence $\widetilde{b_{n,t}} = 0$ and $k_{n,t} = \widetilde{a_{n,t}}$) we see that $\widetilde{a_{n,t}}$ converges in $\mathcal{S}'(\mathbb{R}^{2d})$ as well, hence $\widetilde{b_{n,t}}$ converges in $\mathcal{S}'(\mathbb{R}^{2d})$ by difference. Therefore, for any non-zero $\Psi \in \mathcal{S}(\mathbb{R}^{2d})$ the functions $\sigma_\Psi \widetilde{a_{n,t}}$ and $\sigma_\Psi \widetilde{b_{n,t}}$ converge pointwise in $\mathbb{R}^{4d}$.

For any fixed $z \in \mathbb{R}^{2d}$, set $F_n(\zeta) = \sigma_\Psi \widetilde{a_{n,t}}(z,\zeta)$ and $G_n(\zeta) = \sigma_\Psi \widetilde{b_{n,t}}(z,\zeta)$. By Lemma 4.4.5 and (6.20) we have

$$\sup_{\zeta \in \mathbb{R}^{2d}} \langle\zeta\rangle^s |F_n(\zeta)| \lesssim \|\widetilde{a_{n,t}}\|_{M_{0,s}^\infty} \lesssim \|a_{n,t}\|_{M_{0,s}^\infty} \leq C(t,\epsilon).$$

Similarly, by Lemma 4.4.5 and (6.21),

$$\|G_n\|_{L^1} \lesssim \|\widetilde{b_{n,t}}\|_{M^{\infty,1}} \lesssim \|b_{n,t}\|_{M^{\infty,1}} \leq \epsilon\, C(t).$$

The estimates just obtained have two relevant implications. On the one hand, the first claim of Theorem 6.7.1 is actually proved. On the other hand, the assumptions of Lemma 6.2.6 are satisfied: we have $(F_n + G_n)(\zeta) = \sigma_\Psi k_{n,t}(z,\zeta)$ and $(F + G)(\zeta) = \sigma_\Psi k_t(z,\zeta)$, and therefore we obtain

$$\limsup_{n\to\infty} \left\|\mathcal{F}\left[(k_{n,t} - k_t)\,\overline{T_z\Psi}\right]\right\|_{L^1} \leq 2\epsilon\, C(t).$$

Since ϵ can be made arbitrarily small and the left-hand side is independent of ϵ, we conclude that

$$\lim_{n \to \infty} \left\| \mathcal{F}\left[\left(k_{n,t} - k_t \right) \overline{T_z \Psi} \right] \right\|_{L^1} = 0,$$

in particular $k_{n,t} \to k_t$ in $(\mathcal{F}L^1(\mathbb{R}^{2d}))_{\text{loc}}$.

Finally, by means of a suitable bump function Θ as in the preceding section, for any fixed $z \in \mathbb{R}^{2d}$ we infer

$$\left\| \mathcal{F}\left[\left(e_{n,t} - u_t \right) \overline{T_z \Psi} \right] \right\|_{L^1} \leq |\det B_t|^{-1/2} \left\| \mathcal{F}\left[\left(T_z \Theta e^{2\pi i \Phi_t} \right) \right] \right\|_{L^1} \left\| \mathcal{F}\left[\left(k_{n,t} - k_t \right) \overline{T_z \Psi} \right] \right\|_{L^1},$$

and thus

$$\left\| \mathcal{F}\left[\left(e_{n,t} - u_t \right) \overline{T_z \Psi} \right] \right\|_{L^1} \to 0.$$

This gives $e_{n,t} \to u_t$ in $(\mathcal{F}L^1(\mathbb{R}^{2d}))_{\text{loc}}$ and therefore uniformly on compact subsets of $\mathbb{R}^{2d}$.   □

6.8   Convergence at Exceptional Times

Let us focus here on the occurrence of a set of exceptional times in Theorems 6.5.1 and 6.7.1. It should be clear from the flow of the arguments that this is not a mere technical workaround to make proofs work—in fact, it is an unavoidable and expected phenomenon from a mathematical point of view. In support of this claim, note that the correspondence between free symplectic matrices and quadratic Fourier transforms as in (6.1) (cf. Proposition 4.3.4 for a precise formulation) can be used to determine the abundance of free matrices in $\mathrm{Sp}(d, \mathbb{R})$. It turns out that not being free is an exceptional feature for a symplectic matrix, in the sense that follows.

Proposition 6.8.1 ([73, Proposition 171]) *The subset* $\mathrm{Sp}_0(d, \mathbb{R})$ *of free symplectic matrices has codimension 1 in* $\mathrm{Sp}(d, \mathbb{R})$ *and therefore measure zero.*

Moreover, it may very well happen that the integral kernel of the evolution operator degenerates into a distribution—examples abound also in fundamental settings. Consider for instance the harmonic oscillator (corresponding to the classical Hamiltonian $\frac{1}{2}|\xi|^2 + \frac{1}{2}|x|^2$ and $\hbar = 1/(2\pi)$):

$$\frac{i}{2\pi} \partial_t \psi = -\frac{1}{8\pi^2} \Delta \psi + \frac{1}{2}|x|^2 \psi.$$

The integral kernel of the corresponding evolution operator was computed in Example 4.3.10 and shows the expected degenerate behaviour at integer multiples of π, which is consistent with the fact that the associated classical flow S_t is given by

$$S_t = \begin{bmatrix} (\cos t)I & (\sin t)I \\ -(\sin t)I & (\cos t)I \end{bmatrix};$$

hence $\mathfrak{E} = \{t \in \mathbb{R} : \sin t = 0\} = \{k\pi : k \in \mathbb{Z}\}$.

It is natural to wonder whether convergence of integral kernels still occurs in some distributional sense, hopefully better than the broadest one (that is $\mathcal{S}'(\mathbb{R}^{2d})$). It turn out that a suitable framework is offered by the Banach-Gelfand triple (M^1, L^2, M^∞) of modulation spaces, which has better properties than the standard triple $(\mathcal{S}, L^2, \mathcal{S}')$ of real harmonic analysis (see Sect. 3.4 and also [98] in this connection).

Let us commence the investigation on convergence results for integral kernels at exceptional times with a general result for the kernels of strongly convergent sequences of operators in L^2.

Theorem 6.8.2 *Let $\{A_n\}_{n\in\mathbb{N}} \subset \mathcal{L}(L^2(\mathbb{R}^d))$, be a sequence of bounded linear operators on $L^2(\mathbb{R}^d)$ with associated distribution kernels $\{a_n\}_{n\in\mathbb{N}} \subset \mathcal{S}'(\mathbb{R}^{2d})$, and $A \in \mathcal{L}(L^2(\mathbb{R}^d))$ with distribution kernel $a \in \mathcal{S}'(\mathbb{R}^{2d})$. Assume that $A_n \to A$ in the strong operator topology. Then:*

(i) a_n, $n \in \mathbb{N}$, and a belong to a bounded subset of $M^\infty(\mathbb{R}^{2d})$;

(ii) $a_n \to a$ in the weak-* topology on $M^\infty(\mathbb{R}^{2d})$, that is $\langle a_n, \varphi \rangle \to \langle a, \varphi \rangle$ for all $\varphi \in M^1(\mathbb{R}^{2d})$.

In particular, if we regard a_n as elements of $(\mathcal{F}L^\infty(\mathbb{R}^{2d}))_{\mathrm{loc}}$ then $\langle a_n, \varphi \rangle \to 0$ for every $\varphi \in (\mathcal{F}L^1(\mathbb{R}^{2d}))_{\mathrm{comp}}$.

Proof We have that $\{A_n\}$ is a bounded sequence in $\mathcal{L}(L^2(\mathbb{R}^d))$ as a consequence of the uniform boundedness principle, hence also in $\mathcal{L}(M^1(\mathbb{R}^d), M^\infty(\mathbb{R}^d))$. The Feichtinger kernel theorem (Theorem 3.4.3) yields that the kernels a_n belong to a bounded subset of $M^\infty(\mathbb{R}^{2d})$. Similarly, $A \in \mathcal{L}(L^2(\mathbb{R}^d)) \Rightarrow a \in M^\infty(\mathbb{R}^{2d})$.

For the second part of the claim we remark that $A_n \to A$ in the strong operator topology implies that $a_n \to a$ in $\mathcal{S}'(\mathbb{R}^{2d})$ by Proposition 6.2.1. Therefore, for any fixed non-zero $g \in \mathcal{S}(\mathbb{R}^d)$ we have $V_g a_n \to V_g a$ pointwise in $\mathbb{R}^{2d}$. Moreover, we have the estimate $|V_g a_n(x, \xi)| \le C$, for some constant $C > 0$ independent of n by the first part of the claim. Hence, for any $\varphi \in M^1(\mathbb{R}^{2d})$ we have

$$\langle a_n, \varphi \rangle = \int_{\mathbb{R}^{2d}} V_g a_n(x, \xi) \overline{V_g \varphi(x, \xi)} dx d\xi$$

$$\to \int_{\mathbb{R}^{2d}} V_g a(x, \xi) \overline{V_g \varphi(x, \xi)} dx d\xi = \langle a, \varphi \rangle,$$

by the dominated convergence theorem. $\qquad\qquad\qquad\qquad\qquad\qquad\qquad\square$

As a straightforward application of this result we can prove global-in-time convergence of integral kernels, although in a weaker sense than before.

Corollary 6.8.3 *Assume $H_0 = Q^{\mathrm{w},\hbar}$ as in the summary and $V \in \mathcal{L}(L^2(\mathbb{R}^d))$. Let $e_{n,t}(\hbar, \cdot) \in \mathcal{S}'(\mathbb{R}^{2d})$ be the distribution kernel of the Feynman-Trotter parametrix $E_n(t)$ in (6.5) and $u_t(\hbar, \cdot) \in \mathcal{S}'(\mathbb{R}^{2d})$ be the kernel of the Schrödinger evolution operator $U(t)$ associated with the Cauchy problem (6.4). For any $n \in \mathbb{N}, n \geq 1, t \in \mathbb{R}, \hbar > 0$ we have $e_{n,t}(\hbar, \cdot), u_t(\hbar, \cdot) \in M^\infty(\mathbb{R}^{2d})$. Moreover, $e_{n,t}(\hbar, \cdot) \to u_t(\hbar, \cdot)$ in the weak-$*$ topology on $M^\infty(\mathbb{R}^{2d})$ for any fixed $t \in \mathbb{R}$ and $\hbar > 0$.*

For more regular potentials we expect that the conclusion of Corollary 6.8.3 can be improved. Indeed, the following result can be thought of as a version of the Trotter formula for potentials in $M^{\infty,1}(\mathbb{R}^d)$, with strong convergence on $M^1(\mathbb{R}^d)$.

Theorem 6.8.4 *Assume $H_0 = Q^{\mathrm{w},\hbar}$ and $V = \sigma^{\mathrm{w},\hbar}$ for some $\sigma \in M^{\infty,1}(\mathbb{R}^{2d})$. Let $\{E_n(t)\}, n \geq 1$, be the sequence of Feynman-Trotter parametrices defined in (6.5) and $U(t)$ be the Schrödinger evolution operator $U(t)$ associated with the Cauchy problem (6.4). For any fixed $t \in \mathbb{R}, \hbar > 0$ we have*

$$\lim_{n\to\infty} E_n(t) = U(t), \qquad \lim_{n\to\infty} E_n(t)^* = U(t)^*$$

in the strong topology of operators acting on $M^1(\mathbb{R}^d)$. In particular, for all $t \in \mathbb{R}, 0 < \hbar \leq 1$, and $\varphi \in M^1(\mathbb{R}^d)$, the functions

$$\langle e_{n,t}(\hbar, x, \cdot), \varphi \rangle, \quad \langle e_{n,t}(\hbar, \cdot, y), \varphi \rangle, \quad \langle u_t(\hbar, x, \cdot), \varphi \rangle, \quad \langle u_t(\hbar, \cdot, y), \varphi \rangle$$

belong to $M^1(\mathbb{R}^d)$, and

$$\langle e_{n,t}(\hbar, x, \cdot), \varphi \rangle \to \langle u_t(\hbar, x, \cdot), \varphi \rangle, \quad \langle e_{n,t}(\hbar, \cdot, y), \varphi \rangle \to \langle u_t(\hbar, \cdot, y), \varphi \rangle$$

in $M^1(\mathbb{R}^d)$, hence in $L^p(\mathbb{R}^d)$ for every $1 \leq p \leq \infty$.

Proof We prove that $E_n(t) \to U(t)$ strongly in $\mathcal{L}(M^1(\mathbb{R}^d))$; the claim concerning adjoint operators follows by similar arguments since $U(t)^* = U(-t)$ and $E_n(t)^* = \left(e^{\frac{i}{\hbar}\frac{t}{n}V} e^{\frac{i}{\hbar}\frac{t}{n}H_0} \right)^n$.

As already observed, we know that the operator $H_0 = Q^{\mathrm{w}}$ with domain $D(H_0) = \{f \in L^2(\mathbb{R}^d) : H_0 f \in L^2(\mathbb{R}^d)\}$ is self-adjoint [142]. Let $U_0(t) = e^{-i\frac{t}{\hbar}H_0}$ be the corresponding strongly continuous unitary group on $L^2(\mathbb{R}^d)$. The well-posedness of the Schrödinger equation $i\hbar\partial_t\psi = H_0\psi$ in $M^1(\mathbb{R}^d)$ (see e.g. [58]) implies that the restriction of $U_0(t)$ to $M^1(\mathbb{R}^d)$ defines a strongly continuous group on $M^1(\mathbb{R}^d)$, its generator being the restriction of H_0 to the subspace $\{f \in M^1(\mathbb{R}^d) : H_0 f \in M^1(\mathbb{R}^d)\}$, as a consequence of known results on subspace semigroups, cf. [77, Chapter 2, Section 2.3]. Since a Weyl operator with symbol in the Sjöstrand class is a bounded operator on $M^1(\mathbb{R}^d)$ by Theorem 4.2.2, the desired result follows from the classical Trotter theorem on Banach spaces [77, Corollary 2.7 and Exercise 2.9]. The second part of the claim is just an equivalent formulation of the previous results for the corresponding integral kernels, whereas the last conclusion follows from the continuous embedding $M^1(\mathbb{R}^d) \hookrightarrow L^p(\mathbb{R}^d)$, for every $1 \leq p \leq \infty$. $\qquad\square$

Remark 6.8.5 We expect that the conclusions of Corollary 6.8.3 can be improved in the case where $A_n = E_n(t)$, $A = U(t)$. For instance, convergence results for the corresponding integral kernels could be further explored in the framework of mixed modulation spaces, for which generalized kernel theorems were proved in [49].

6.9 Physics at Exceptional Times

Throughout the previous section we focused on the mathematical nature of exceptional times and the partial extensions of the main convergence results proved before. However, the physics behind exceptional times is not clear at the moment. In fact, this seems to be a highly non-trivial question—although, curiously enough, it appears as an exercise in the textbook [104, Problem 3-1] by Feynman and Hibbs. While heuristic arguments relying on dimensional analysis could provide some hints towards the answer, we attempt to provide a more quantitative solution in terms of measurable quantities.

Let $B(u, r)$ be the (open) ball with center $u \in \mathbb{R}^d$ and radius $r > 0$ in $\mathbb{R}^d$. In accordance with the custom in physics we temporarily adopt the bra-ket notation and identify quantum states with the corresponding wave functions in the position representation.

Fix $x_0, y_0 \in \mathbb{R}^d$ and $a, b > 0$, and consider the normalized wave packets

$$|A\rangle = \frac{1}{\sqrt{|B(y_0, a)|}} 1_{B(y_0, a)}, \quad |B\rangle = \frac{1}{\sqrt{|B(x_0, b)|}} 1_{B(x_0, b)}.$$

The corresponding transition amplitude from the state $|A\rangle$ to $|B\rangle$ under the Hamiltonian $H = H_0 + V$ as in Theorem 6.7.1, namely

$$I = I(t, x_0, y_0, a, b) = \langle B|U(t)|A\rangle, \quad t \in \mathbb{R},$$

trivially satisfies the estimate

$$|I(t, x_0, y_0, a, b)| \leq 1, \quad \forall t \in \mathbb{R}, \ x_0, y_0 \in \mathbb{R}^d, \ a, b > 0.$$

It is easy to realize that this bound cannot be improved at exceptional times—consider for instance the case where $t = 0$, $x_0 = y_0$ and $a = b$, which yields $I = 1$. Nevertheless, we have the following result.

Proposition 6.9.1 *Under the same assumptions of Theorem 6.7.1, for all $t \in \mathbb{R} \setminus \mathfrak{E}$ and $x_0, y_0 \in \mathbb{R}^d$ we have*

$$\lim_{a,b \to 0} \frac{I(t, x_0, y_0, a, b)}{(ab)^{d/2}} = C \overline{u_t(\hbar, x_0, y_0)},$$

where $C = C(d) = |B(0, 1)|$.

Proof An explicit computation yields

$$\frac{I(t, x_0, y_0, a, b)}{C(ab)^{d/2}} = \frac{1}{C^2(ab)^d} \int_{B(x_0,b)} \int_{B(y_0,a)} \overline{u_t(\hbar, x, y)} \, dy \, dx,$$

and the conclusion follows by the continuity of $u_t(\hbar, x, y)$ in $\mathbb{R}^{2d}$, because $u_t(\hbar, \cdot) \in (\mathcal{F}L^1(\mathbb{R}^{2d}))_{\mathrm{loc}}$ for $t \in \mathbb{R} \setminus \mathfrak{E}$ by Theorem 6.7.1.   □

This result shows that while $|I| \leq 1$ in general, for a non-exceptional time $t \in \mathbb{R} \setminus \mathfrak{E}$ we have that $|I| \sim (ab)^{d/2}$ as $a, b \to 0$. In particular $|I| \to 0$ as $a, b \to 0$ except (possibly) for exceptional times.

Chapter 7
Convergence in $\mathcal{L}(L^2)$ for Potentials in the Sjöstrand Class

7.1 Summary

The main motivation for both this chapter and the subsequent one comes from a deeper analysis of Fujiwara's convergence result in $\mathcal{L}(L^2)$ for time-slicing approximations of Feynman path integrals (cf. Theorem 1.3.1). As already remarked in Sect. 1.3.1, there is good reason to believe that the assumptions on the regularity of the potential may be relaxed in order to preserve convergence in operator norm. In this chapter we are concerned with the equation

$$i\hbar\partial_t \psi = -\frac{1}{2}\hbar^2\Delta\psi + V(t,x)\psi,$$

for a potential V satisfying the following condition.

Assumption ($\tilde{\mathbf{A}}$) *V is a real-valued function of $(t,x) \in \mathbb{R} \times \mathbb{R}^d$ satisfying,[1] for some $N \in \mathbb{N}$, $N \geq 1$,*

$$\partial_t^k \partial_x^\alpha V \in C_b(\mathbb{R}; M^{\infty,1}(\mathbb{R}^d)),$$

for any $k \in \mathbb{N}$ and $\alpha \in \mathbb{N}^d$ satisfying $2k + |\alpha| \leq 2N$.

Roughly speaking, potentials satisfying Assumption ($\tilde{\text{A}}$) are bounded continuous functions together with a certain number of derivatives. Assumptions in the same spirit, or even stronger (e.g., smooth potentials with compact support), are quite popular in scattering theory [176].

We will construct parametrices different from those in (1.18), motivated by the following remarks. First, note that in Fujiwara's estimates the approximation power

---

[1] $C_b(\mathbb{R}; X)$ is the space of bounded continuous functions $f: \mathbb{R} \to X$; see Chap. 2.

© The Author(s), under exclusive license to Springer Nature Switzerland AG 2022
F. Nicola, S. I. Trapasso, *Wave Packet Analysis of Feynman Path Integrals*,
Lecture Notes in Mathematics 2305, https://doi.org/10.1007/978-3-031-06186-8_7

increases with N from the point of view of semiclassical analysis (i.e., positive powers of $\hbar$ on the right-hand side of (1.20)). Nevertheless, such parametrices have limited applicability to concrete situations and computational problems, the main obstruction being the required knowledge of the exact action functional—generally unattainable, except for quite simple systems. In light of this difficulty it is beneficial to consider short-time approximations for the action obtained by means of the so-called *midpoint rules* [201]. For example, in the case of a standard Hamiltonian $H(x, \xi) = |\xi|^2/2 + V(x)$ the action is

$$S(t, s, x, y) = \frac{|x - y|^2}{2(t - s)} - \mathcal{V}(t, s, x, y), \quad \mathcal{V} = \int_s^t V(\gamma(\tau))d\tau,$$

the latter integral involving a path γ with $\gamma(s) = y$ and $\gamma(t) = x$. Midpoint rules essentially amount to replace $\mathcal{V}$ with approximate expressions like

$$\mathcal{V}_1 = \frac{V(x) + V(y)}{2}(t - s), \quad \text{or} \quad \mathcal{V}_2 = V\left(\frac{x + y}{2}\right)(t - s).$$

Despite their popularity within the physics literature, it is not difficult to realize that these rules are not sufficiently accurate. Consider a simple test for the harmonic oscillator; the corresponding approximate actions S_1, S_2 satisfy

$$S(t, s, x, y) - S_j(t, s, x, y) = O(t - s), \quad j = 1, 2.$$

The quest for accurate, yet manageable, short-time approximations was initiated by N. Makri and W. Miller [169–171], leading to the optimal solution given by the integral average

$$\overline{\mathcal{V}}(s, x, y) = \int_0^1 V(\tau x + (1 - \tau)y, s)d\tau.$$

This precept indeed satisfies a correct first-order approximation, i.e. $S(t, s, x, y) - \overline{S}(t, s, x, y) = O((t - s)^2)$ for small $t - s$. The interested reader can find more details on the issue in the aforementioned papers, as well as the recent one [74] by de Gosson.

Inspired by this discussion, and the current practice in physics and chemistry, we introduce different time-slicing approximation operators than (1.18), namely

$$\widetilde{E}^{(N)}(t, s) f(x) = \frac{1}{(2\pi i \hbar (t - s))^{d/2}} \int_{\mathbb{R}^d} e^{\frac{i}{\hbar} S^{(N)}(t, s, x, y)} f(y)dy, \tag{7.1}$$

where the approximate action $S^{(N)}$ ultimately coincides with a Taylor-like expansion of the exact action S at $t = s$:

$$S^{(N)}(t, s, x, y) = \frac{|x - y|^2}{2(t - s)} + \sum_{k=1}^{N} W_k(s, x, y)(t - s)^k. \tag{7.2}$$

The coefficients $W_k(s, x, y)$ are recursively constructed in light of a careful analysis of power series solutions for the following modified Hamilton-Jacobi equation

$$\frac{\partial S}{\partial t} + \frac{1}{2}|\nabla_x S|^2 + V(t, x) + \frac{i\hbar d}{2(t - s)} - \frac{i\hbar}{2}\Delta_x S = 0.$$

In particular, the last two non-standard terms are precisely tailored to boost the approximation power of the parametrix $\widetilde{E}^{(N)}$. In any case, the "counterterm" is first order in $\hbar$ and identically vanishes in the free particle case ($V = 0$). Moreover, we remark that $W_1(s, x, y) = \mathcal{V}(s, x, y)$ as expected. All these aspects are discussed in detail in Sect. 7.3, which is devoted to a rigorous short-time analysis of the action functional.

Given a subdivision $\Omega = \{t_0, \ldots, t_L\}$ of the interval $[s, t]$ such that $s = t_0 < t_1 < \ldots < t_L = t$, we introduce the long-time composition

$$\widetilde{E}^{(N)}(\Omega, t, s) = \widetilde{E}^{(N)}(t_L, t_{L-1})\widetilde{E}^{(N)}(t_{L-1}, t_{L-2}) \cdots \widetilde{E}^{(N)}(t_1, t_0),$$

which has integral kernel

$$K^{(N)}(\Omega, t, s, x, y) = \prod_{j=1}^{L} \frac{1}{(2\pi i(t_j - t_{j-1})\hbar)^{d/2}}$$

$$\times \int_{\mathbb{R}^{d(L-1)}} \exp\left(\frac{i}{\hbar}\sum_{j=1}^{L} S^{(N)}(t_j, t_{j-1}, x_j, x_{j-1})\right) \prod_{j=1}^{L-1} dx_j,$$

with $x = x_L$ and $y = x_0$. It is reasonable to believe that the operators $\widetilde{E}^{(N)}(\Omega, t, s)$ converge to the actual propagator as $\omega(\Omega) = \sup\{t_j - t_{j-1} : j = 1, \ldots, L\} \to 0$, in line with Feynman's insight. This is precisely the content of Theorem 7.4.4, where convergence in $\mathcal{L}(L^2)$ for these rough time-slicing approximations is proved under Assumption $(\widetilde{A})$ for the potential. The proof of this result can be found in Sect. 7.4, after a careful analysis of the properties satisfied by the parametrices $\widetilde{E}^{(N)}(t, s)$. We emphasize that the proof relies on a general approximation scheme that is proved separately in Sect. 7.2.

7.2 An Abstract Approximation Result in $\mathcal{L}(L^2)$

We begin the journey towards Theorem 7.4.4 by providing an abstract approximation result that will be invoked in different occasions below. It can be regarded as a
generalization of [109, Lemma 3.2].

Theorem 7.2.1 *Assume that for some $\delta > 0$ we have a family of operators $E(t, s)$
for $0 < t - s \leq \delta$, and $U(t, s)$, $s, t \in \mathbb{R}$, bounded in $L^2(\mathbb{R}^d)$, satisfying the following
conditions:*

*(i) U enjoys the evolution property $U(t, \tau)U(\tau, s) = U(t, s)$ for every $s < \tau < t$
and for every $T > 0$ there exists a constant $C_0 \geq 1$ such that*

$$\|U(t, s)\|_{L^2 \to L^2} \leq C_0 \quad for \quad 0 < t - s \leq T. \tag{7.3}$$

(ii) There exists $C_1 > 0$ such that

$$\|E(t, s) - U(t, s)\|_{L^2 \to L^2} \leq C_1(t - s)^{N+1} \quad for \quad 0 < t - s \leq \delta. \tag{7.4}$$

*For any subdivision $\Omega : s = t_0 < t_1 < \ldots < t_L = t$ of the interval $[s, t]$, with
$\omega(\Omega) = \sup\{t_j - t_{j-1} : j = 1, \ldots, L\} < \delta$, consider therefore the composition
$E(\Omega, t, s)$ defined by*

$$E(\Omega, t, s) = E(t_L, t_{L-1})E(t_{L-1}, t_{L-2}) \cdots E(t_1, t_0),$$

Then, for every $T > 0$ there exists a constant $C = C(T) > 0$ such that

$$\|E(\Omega, t, s) - U(t, s)\|_{L^2 \to L^2} \leq C\omega(\Omega)^N (t - s) \tag{7.5}$$

for $0 < t - s \leq T$. More precisely,

$$C = C(T) = C_0^2 C_1 \exp\left(C_0 C_1 \omega(\Omega)^N T\right).$$

Proof Set

$$R(t, s) := E(t, s) - U(t, s)$$

so that by (7.4) we have

$$\|R(t, s)\|_{L^2 \to L^2} \leq C_1(t - s)^{N+1} \text{ for } 0 < t - s \leq \delta. \tag{7.6}$$

Hence we can write

$$E(\Omega, t, s) - U(t, s) = \big(U(t, t_{L-1}) + R(t, t_{L-1})\big) \dots \big(U(t_1, s) + R(t_1, s)\big) - U(t, s).$$

One expands the product above and obtains a sum of ordered products of operators, where each product has the following structure: *from right to left* we have, say, q_1 factors of type U, p_1 factors of type R, q_2 factors of type U, p_2 factors of type R, etc., up to q_k factors of type U, p_k factors of type R, to finish with q_{k+1} factors of type U. We can schematically write such a product as

$$\underbrace{U \dots U}_{q_{k+1}} \underbrace{R \dots R}_{p_k} \underbrace{U \dots U}_{q_k} \dots \dots \underbrace{R \dots R}_{p_1} \underbrace{U \dots U}_{q_1}.$$

Here $p_1, \dots, p_k, q_1, \dots q_k, q_{k+1}$ are non-negative integers whose sum is L, with $p_j > 0$ and we can of course group together the consecutive factors of type U, using the evolution property assumed for U. Now, for $0 < t - s \le T$ we estimate the $L^2 \to L^2$ norm of the above ordered product using the known estimates for each factor, namely (7.3) and (7.6). In particular, by using the assumption $C_0 \ge 1$, we get

$$\|E(\Omega, t, s) - U(t, s)\|_{L^2 \to L^2} \le C_0^{k+1} \prod_{j=1}^{k} \prod_{i=1}^{p_j} C_1 (t_{J_j+i} - t_{J_j+i-1})^{N+1}$$

$$\le C_0 \prod_{j=1}^{k} \prod_{i=1}^{p_j} C_0 C_1 (t_{J_j+i} - t_{J_j+i-1})^{N+1},$$

where $J_j = p_1 + \dots + p_{j-1} + q_1 + \dots + q_j$ for $j \ge 2$ and $J_1 = q_1$. The sum over $p_1, \dots, p_k, q_1, \dots, q_{k+1}$ of these terms gives in turn

$$\|E(\Omega, t, s) - U(t, s)\|_{L^2 \to L^2} \le C_0 \left\{ \prod_{j=1}^{L} (1 + C_0 C_1 (t_j - t_{j-1})^{N+1}) - 1 \right\}$$

$$\le C_0 \left\{ \exp\left(\sum_{j=1}^{L} C_0 C_1 (t_j - t_{j-1})^{N+1} \right) - 1 \right\}$$

$$\le C_0 \left\{ \exp\left(C_0 C_1 \omega(\Omega)^N (t - s) \right) - 1 \right\}$$

$$\le C_0^2 C_1 \omega(\Omega)^N (t - s) \exp\left(C_0 C_1 \omega(\Omega)^N (t - s) \right),$$

where in the last inequality we used $e^\tau - 1 \le \tau e^\tau$, for $\tau \ge 0$. This gives (7.5) with $C = C(T)$ as in the statement and concludes the proof.   □

7.3   Short-Time Analysis of the Action

Let us commence with a brief motivational discussion devoted to explain the structure of formula (7.2) for the approximate action $S^{(N)}(t, s, x, y)$. We refer to [73, Section 4.5] for more details.

It is well known (see [104, Section 2.1] and also [80, 166]) that for a classical Hamiltonian

$$H(x, \xi, t) = \frac{1}{2}|\xi|^2 + V(t, x),$$

the action $S(t, s, x, y)$ satisfies the Hamilton-Jacobi equation

$$\frac{\partial S}{\partial t} + \frac{1}{2}|\nabla_x S|^2 + V(t, x) = 0.$$

In order for $\widetilde{E}^{(N)}$ to be a parametrix in a sense to be specified later (cf. (7.13) and (7.14) below), we introduce the slightly modified equation

$$\frac{\partial S}{\partial t} + \frac{1}{2}|\nabla_x S|^2 + V(t, x) + \frac{i\hbar d}{2(t - s)} - \frac{i\hbar}{2}\Delta_x S = 0,$$

and look for a solution S in the form $S(t, s, x, y) = \frac{|x-y|^2}{2(t-s)} + R(t, s, x, y)$, $s < t$. This yields an equivalent equation for R, namely

$$\frac{\partial R}{\partial t} + \frac{1}{2}|\nabla_x R|^2 + V(t, x) + \frac{1}{t - s}(x - y) \cdot \nabla_x R - \frac{i\hbar}{2}\Delta_x R = 0.$$

Assume that

$$R(t, s, x, y) = W_0 + W_1(s, x, y)(t - s) + W_2(s, x, y)(t - s)^2 + \dots,$$

where the functions $W_k(s, x, y)$ will be briefly denoted by $W_k(x, y)$ from now on. We immediately find $W_0 = 0$ and, for $k \geq 1$, by equating to 0 the coefficient of the term $(t - s)^{k-1}$, we obtain the equations

$$kW_k(x, y) + (x - y) \cdot \nabla_x W_k(x, y) = F_k(s, x, y), \tag{7.7}$$

where we set

$$F_k(s, x, y) = -\frac{1}{2}\sum_{\substack{j+\ell=k-1 \\ j\geq 1, \ell\geq 1}} \nabla_x W_j \cdot \nabla_x W_\ell - \frac{1}{(k-1)!}\partial_t^{k-1}V(s, x) + \frac{i\hbar}{2}\Delta_x W_{k-1}.$$

$$\tag{7.8}$$

For brevity we also write $F_k(x, y)$ in place of $F_k(s, x, y)$.

Lemma 7.3.1 *Suppose F_k in (7.7) is continuous as a function of $(x, y) \in \mathbb{R}^{2d}$. Then there exists a unique continuous solution of (7.7), namely*

$$W_k(x, y) = \int_0^1 \tau^{k-1} F_k(\tau x + (1 - \tau)y, y)d\tau. \tag{7.9}$$

Proof The argument relies on the methods of characteristics. To be precise, the original PDE (7.7) reduces to a linear ODE with respect to the variable $\lambda \in \mathbb{R}$ along the curves of type $x_u(\lambda) = y + ue^\lambda$ for unit-norm $u \in \mathbb{R}^d$, namely

$$\frac{d}{d\lambda} W_k(x_u(\lambda), y) + k W_k(x_u(\lambda), y) = F_k(x_u(\lambda), y).$$

The solutions of the latter ODE are given by

$$W_k(x_u(\lambda), y) = e^{-k\lambda} \left(\int_{-\infty}^\lambda e^{k\sigma} F_k(x_u(\sigma), y)d\sigma + C \right),$$

where $C \in \mathbb{R}$ is an arbitrary constant. Notice that $\lambda = \log |x - y|$ and the change of variable $\sigma = \log(|x - y|\tau)$ thus gives

$$W_k(x, y) = \int_0^1 \tau^{k-1} F_k(\tau x + (1 - \tau)y, y)d\tau + \frac{C}{|x - y|^k}.$$

It is then evident that the unique continuous solution corresponds to $C = 0$. □

We now assume that V satisfies Assumption ($\tilde{\mathrm{A}}$) and we prove that we can then solve the Eq. (7.7) for $k = 1, \ldots, N$ by applying repeatedly Lemma 7.3.1 above.

Proposition 7.3.2 *Let V satisfy Assumption ($\tilde{\mathrm{A}}$). Then Eq. (7.7) has, for any $1 \le k \le N$, a unique solution $W_k(s, x, y)$ satisfying*

$$\|\partial_x^\alpha W_k\|_{M^{\infty,1}(\mathbb{R}^{2d})} \le C, \quad \text{for } |\alpha| \le 2(N - k + 1), \ s \in \mathbb{R},$$

for some constant $C > 0$.

Proof First of all, we recall that any function in $M^{\infty,1}$ is continuous. Let us first prove the claim for $k = 1$. We have $F_1(s, x, y) = -V(s, x)$. Using Lemma 7.3.1 with $k = 1$, the STFT of $\partial_x^\alpha W_1(s, \cdot, \cdot)$, $|\alpha| \le 2N$, can be written as

$$\left| V_g \partial_x^\alpha W_1(z, \zeta) \right| = \left| \int_0^1 \tau^{|\alpha|} V_g \left[\partial_x^\alpha V(s, \tau x + (1 - \tau)y) \right](z, \zeta) d\tau \right|, \quad z, \zeta \in \mathbb{R}^{2d}.$$

We now think of V as a function on $\mathbb{R}^{2d}$. More precisely, define

$$V'(s, x, y) = V(s, x), \quad s \in \mathbb{R}, \ x, y \in \mathbb{R}^d,$$

and notice that V' still satisfies Assumption $(\tilde{A})$ with $M^{\infty,1}(\mathbb{R}^d)$ replaced by $M^{\infty,1}(\mathbb{R}^{2d})$.

Let us introduce the parametrized matrices $M_\tau = \begin{bmatrix} \tau I & (1-\tau)\,I \\ 0 & I \end{bmatrix} \in \mathrm{GL}(2d,\mathbb{R})$,

with $\tau \in (0,1]$. We can thus write $V(s, \tau x + (1-\tau)\,y) = V'(s, M_\tau(x,y))$, and by the behaviour of modulation spaces under dilations (cf. Proposition 3.2.4) we have $\partial_x^\alpha V'(s, M_\tau(x,y)) \in M^{\infty,1}(\mathbb{R}^{2d})$. Therefore,

$$\left\| \partial_x^\alpha W_1 \right\|_{M^{\infty,1}} \lesssim \int_0^1 \tau^{|\alpha|} \left\| \partial_x^\alpha V'(s, M_\tau \cdot) \right\|_{M^{\infty,1}} d\tau$$

$$\lesssim \left(\int_0^1 \tau^{|\alpha|} C_{\infty,1}(M_\tau) d\tau \right) \left\| \partial_x^\alpha V' \right\|_{M^{\infty,1}} \leq C,$$

where

$$C_{\infty,1}(M_\tau) = \left(\det(I + M_\tau^\top M_\tau) \right)^{1/2}$$

is a continuous (hence bounded) function of the parameter $\tau \in [0,1]$.

Assume now that the claim holds for any W_j up to a certain $k \leq N-1$ and consider

$$|V_g \partial_x^\alpha W_{k+1}(z,\zeta)| = \left| \int_0^1 \tau^{k+|\alpha|} V_g[\partial_x^\alpha F_{k+1}(\tau x + (1-\tau)y, y)](z,\zeta) d\tau \right|.$$

It is easy to deduce from (7.8) and the hypothesis on $W_1, \ldots, W_k$ that $\partial_x^\alpha F_{k+1}(x,y) \in M^{\infty,1}(\mathbb{R}^{2d})$ whenever $|\alpha| \leq 2(N-k)$. Again by Proposition 3.2.4 we have $\partial_x^\alpha F_{k+1}(M_\tau(x,y)) \in M^{\infty,1}(\mathbb{R}^{2d})$, and arguing as before we have

$$\left\| \partial_x^\alpha W_{k+1} \right\|_{M^{\infty,1}} \lesssim \int_0^1 \tau^{k+|\alpha|} \left\| \partial_x^\alpha F_{k+1}(M_\tau \cdot) \right\|_{M^{\infty,1}} d\tau$$

$$\lesssim \left(\int_0^1 \tau^{k+|\alpha|} C_{\infty,1}(M_\tau) d\tau \right) \left\| \partial_x^\alpha W_{k+1} \right\|_{M^{\infty,1}}$$

$$\leq C.$$

The claim is then proved by induction.   □

Let us consider now the approximate generating functions introduced in (7.2), namely

$$S^{(N)}(t,s,x,y) = \frac{|x-y|^2}{2(t-s)} + R^{(N)}(t,s,x,y),$$

where

$$R^{(N)}(t, s, x, y) := \sum_{k=1}^{N} W_k(x, y)(t - s)^k \qquad (7.10)$$

and $W_k(x, y)$ is defined in (7.9). It is worth to highlight that the first-order approximation of the action is thus given by

$$S^{(1)}(t, s, x, y) = \frac{|x - y|^2}{2(t - s)} - (t - s) \int_0^1 V(s, \tau x + (1 - \tau)y)d\tau.$$

We conclude this section with a uniform estimate for $e^{\frac{i}{\hbar} R^{(N)}}$ in $M^{\infty,1}$ that will be used below.

Proposition 7.3.3 *If the potential function V satisfies Assumption (Ã), then $e^{\frac{i}{\hbar} R^{(N)}} \in M^{\infty,1}(\mathbb{R}^{2d})$, with $R^{(N)}$ as in (7.10). More precisely,*

$$\|e^{\frac{i}{\hbar} R^{(N)}}\|_{M^{\infty,1}} \le C(T),$$

for $0 \le t - s \le T\hbar, 0 < \hbar \le 1$.

Proof If V satisfies Assumption (Ã), Proposition 7.3.2 holds and $\partial^\alpha W_k(x, y) \in M^{\infty,1}(\mathbb{R}^{2d})$ for any $|\alpha| \le 2(N - k + 1)$. In particular, $W_k \in M^{\infty,1}(\mathbb{R}^{2d})$ for all $k = 1, \ldots, N$ and thus $R^{(N)} \in M^{\infty,1}(\mathbb{R}^{2d})$.

Recall from Sect. 3.5 that $M^{\infty,1}(\mathbb{R}^{2d})$ is a Banach algebra for pointwise multiplication (the normalization is such that the unit element has unit norm), hence it is enough to show the desired estimate for $e^{\frac{i}{\hbar}(t-s)^k W_k}$, for any $1 \le k \le N$. We obtain

$$\left\|e^{\frac{i}{\hbar}(t-s)^k W_k}\right\|_{M^{\infty,1}} = \left\|\sum_{n=0}^{\infty} \frac{i^n (t - s)^{kn} (W_k)^n}{\hbar^n n!}\right\|_{M^{\infty,1}}$$

$$\le \sum_{n=0}^{\infty} \frac{(t - s)^{kn} \|W_k\|_{M^{\infty,1}}^n}{\hbar^n n!}$$

$$= e^{\hbar^{-1}(t-s)^k \|W_k\|_{M^{\infty,1}}}$$

$$= e^{T^k \|W_k\|_{M^{\infty,1}}} \le C(T),$$

where we used $0 \le t - s \le T\hbar, 0 < \hbar \le 1$, and Proposition 7.3.2. $\qquad\square$

7.4   Estimates for the Parametrix and Convergence Results

Let us first recall that the Cauchy problem for the Schrödinger equation with bounded potentials is globally well-posed in $L^2(\mathbb{R}^d)$. Arguing as in the proof of Theorem 6.4.1, one can prove the following basic result.

Proposition 7.4.1 *Assume that V is a real-valued function on $\mathbb{R} \times \mathbb{R}^d$ satisfying $V \in C_b(\mathbb{R}; L^\infty(\mathbb{R}^d))$ and let $s \in \mathbb{R}$. Then, the Cauchy problem*

$$\begin{cases} i\hbar \partial_t \psi = -\frac{1}{2}\hbar^2 \Delta \psi + V(t,x)\psi \\ \psi(s,x) = f(x) \end{cases}$$

is (backward and) forward globally well-posed in $L^2(\mathbb{R}^d)$ and the corresponding propagator $U(t,s)$ is a unitary operator on $L^2(\mathbb{R}^d)$.

Consider now the parametrix $\widetilde{E}^{(N)}(t,s)$ in (7.1). We have the following result.

Proposition 7.4.2 *For every $T > 0$ there exists $C = C(T) > 0$ such that, for $0 < t - s \le T\hbar$, $0 < \hbar \le 1$, we have*

$$\|\widetilde{E}^{(N)}(t,s)\|_{L^2 \to L^2} \le C. \tag{7.11}$$

Moreover, for any $f \in L^2(\mathbb{R}^d)$ we have

$$\lim_{t \searrow s} \widetilde{E}^{(N)}(t,s)f = f. \tag{7.12}$$

Proof First, notice that

$$\widetilde{E}^{(N)}(t,s)f(x) = \frac{1}{(2\pi i (t-s)\hbar)^{d/2}} \int_{\mathbb{R}^d} e^{\frac{i}{\hbar} \frac{|x-y|^2}{2(t-s)}} e^{\frac{i}{\hbar} R^{(N)}(t,s,x,y)} f(y) dy$$

is an oscillatory integral operator (OIO for short) with the free-particle-action as phase function and amplitude (apart from the factor in front of the integral)

$$a^{(N)}(t,s,x,y) := e^{\frac{i}{\hbar} R^{(N)}(t,s,x,y)} \in M^{\infty,1}(\mathbb{R}^{2d})$$

by Proposition 7.3.3. We would like to apply Lemma 4.4.10. To this end, we need some preparation. First, using suitable unitary dilation operators (cf. Sect. 2.3 for notation) we rephrase $\widetilde{E}^{(N)}(t,s)$ as follows:

$$\widetilde{E}^{(N)}(t,s) = D_{\frac{1}{\sqrt{\hbar(t-s)}}} B^{(N)}(t,s) D_{\sqrt{\hbar(t-s)}},$$

where

$$B^{(N)}(t, s) f(x) = \frac{1}{(2\pi i)^{d/2}} \int_{\mathbb{R}^d} e^{\frac{i}{2}|x-y|^2} b^{(N)}(t, s, x, y) f(y) dy$$

is an OIO whose phase function is free from time and $\hbar$ dependence and the amplitude is

$$b^{(N)}(t, s, x, y) = e^{\frac{i}{\hbar} \sum_{k=1}^{N} W_k(\sqrt{\hbar(t-s)}x, \sqrt{\hbar(t-s)}y)(t-s)^k}$$

$$= a^{(N)}(\sqrt{\hbar(t - s)}x, \sqrt{\hbar(t - s)}y).$$

In particular, by Proposition 3.2.4 we infer $b^{(N)} \in M^{\infty,1}(\mathbb{R}^{2d})$ and

$$\|b^{(N)}\|_{M^{\infty,1}} \leq C(T) \|a^{(N)}\|_{M^{\infty,1}}$$

for $0 < \hbar(t - s) \leq T$ (in particular for $0 < t - s \leq \hbar T$, since $0 < \hbar \leq 1$).

Formula (7.11) then will follow from Lemma 4.4.10 and Proposition 7.3.3:

$$\|\widetilde{E}^{(N)}(t, s)\|_{L^2 \to L^2} = \|B^{(N)}(t, s)\|_{L^2 \to L^2}$$

$$\leq C \|b^{(N)}\|_{M^{\infty,1}}$$

$$\leq C(T) \|a^{(N)}\|_{M^{\infty,1}} \leq C'(T),$$

for $0 < t - s \leq T\hbar$.

Let us consider now the issue of strong convergence to the identity as $t \searrow s$. To this aim, we conveniently introduce the operator

$$H^{(N)}(t, s) f(x) = \frac{1}{(2\pi i (t - s) \hbar)^{d/2}} \int_{\mathbb{R}^d} e^{\frac{i}{\hbar} \frac{|x-y|^2}{2(t-s)}} \left(e^{\frac{i}{\hbar} R^{(N)}(t,s,x,y)} - 1 \right) f(y) dy.$$

Using again suitable dilations we obtain

$$H^{(N)}(t, s) = D_{\frac{1}{\sqrt{\hbar(t-s)}}} Q^{(N)}(t, s) D_{\sqrt{\hbar(t-s)}},$$

where

$$Q^{(N)}(t, s) f(x) = \frac{1}{(2\pi i)^{d/2}} \int_{\mathbb{R}^d} e^{\frac{i}{2}|x-y|^2} q^{(N)}(t, s, x, y) f(y) dy$$

is an OIO with amplitude

$$q^{(N)}(t, s, x, y) := b^{(N)}(t, s, x, y) - 1 \in M^{\infty,1}(\mathbb{R}^{2d}).$$

The latter can be expanded as follows:

$$q^{(N)}(t, s, x, y) = e^{\frac{i}{\hbar} R^{(N)}(t,s,\sqrt{\hbar(t-s)}x,\sqrt{\hbar(t-s)}y)} - 1$$

$$= \frac{i}{\hbar}(t-s)\,\overline{R^{(N)}}\left(t, s, \sqrt{\hbar(t-s)}x, \sqrt{\hbar(t-s)}y\right),$$

where

$$\overline{R^{(N)}}(t, s, x, y) = \sum_{n=1}^{\infty} \frac{i^{n-1}}{n!}\left(\frac{t-s}{\hbar}\right)^{n-1}$$

$$\times \left(\sum_{k=1}^{N} W_k\left(\sqrt{\hbar(t-s)}\,x, \sqrt{\hbar(t-s)}\,y\right)(t-s)^{k-1}\right)^n.$$

The Banach algebra property of the Sjöstrand class (cf. Proposition 3.5.3) and the properties of modulation spaces under dilation (cf. Proposition 3.2.4) imply that $\overline{R^{(N)}}$ belongs to a bounded subset of $M^{\infty,1}(\mathbb{R}^{2d})$ for $0 < t - s \le T\hbar, 0 < \hbar \le 1$. It is then clear that $q^{(N)} \to 0$ in $M^{\infty,1}(\mathbb{R}^{2d})$ for $t \searrow s$. Therefore, the OIO $Q^{(N)}$ with amplitude $q^{(N)}$ has operator norm converging to 0 as $t \searrow s$ by Lemma 4.4.10. The same holds for the unitarily equivalent operator $H^{(N)}$.

On the other hand, by the very definition of $H^{(N)}$ we have

$$H^{(N)}(t, s) = \widetilde{E}^{(N)}(t, s) - U_0(t, s),$$

where $U_0(t, s)$ is the free propagator, with strong convergence to the identity operator as $t \searrow s$. Hence (7.12) follows.   □

A direct check shows that $\widetilde{E}^{(N)}(t, s)$ is a parametrix. Precisely we have

$$\left(i\hbar\partial_t + \frac{1}{2}\hbar^2\Delta - V(t, x)\right)\widetilde{E}^{(N)}(t, s) = G^{(N)}(t, s), \tag{7.13}$$

with

$$G^{(N)}(t, s)f = \frac{1}{(2\pi i(t-s)\hbar)^{d/2}}\int_{\mathbb{R}^d} e^{\frac{i}{\hbar}S^{(N)}(t,s,x,y)} g_N(t, s, x, y) f(y)\, dy,$$

$$\tag{7.14}$$

where, from the construction of $S^{(N)}$ (see in particular Eqs. (7.2), (7.7), and (7.8)), the amplitudes g_N is given by

$$g_N(t, s, x, y) = -\frac{\partial S^{(N)}}{\partial t} - \frac{1}{2}\left|\nabla_x S^{(N)}\right|^2 - V(t, x) - \frac{i\hbar d}{2(t-s)} + \frac{i\hbar}{2}\Delta_x S^{(N)}$$

$$= -\frac{1}{2}\sum_{k=N}^{2N}\sum_{\substack{j+\ell=k \\ j,\ell\geq 1}} \nabla_x W_j \cdot \nabla_x W_\ell (t-s)^k$$

$$+ \frac{i\hbar}{2}\Delta_x W_N(x, y)(t-s)^N$$

$$- \frac{(t-s)^N}{(N-1)!}\int_0^1 (1-\tau)^{N-1}\left(\partial_t^N V\right)((1-\tau)s + \tau t, x)\, d\tau.$$

Hence, by Assumption (Ã) and Proposition 7.3.2 we have

$$\|g_N(t, s, \cdot, \cdot)\|_{M^{\infty,1}(\mathbb{R}^{2d})} \leq C(t-s)^N, \tag{7.15}$$

for $0 < t - s \leq T$, with a constant $C = C(T) > 0$ independent of $\hbar \in (0, 1]$.

The preceding discussion is the bedrock of the following result.

Theorem 7.4.3 *For every $T > 0$, there exists a constant $C = C(T) > 0$ such that*

$$\|\widetilde{E}^{(N)}(t, s) - U(t, s)\|_{L^2 \to L^2} \leq C\hbar^{-1}(t-s)^{N+1}, \tag{7.16}$$

whenever $0 < t - s \leq T\hbar$.

Proof We can write the operator $G^{(N)}(t, s)$ in (7.14) as

$$G^{(N)}(t, s)f = \frac{1}{(2\pi i(t-s)\hbar)^{d/2}}\int_{\mathbb{R}^d} e^{\frac{i}{\hbar}\frac{|x-y|^2}{2(t-s)}} e^{\frac{i}{\hbar}R^{(N)}(t,s,x,y)}g_N(t, s, x, y)f(y)\, dy.$$

Following the steps of the proof of Proposition 7.4.2, by means of suitable dilations we can see that $G^{(N)}(t, s)$ is unitarily equivalent to an OIO with phase $|x-y|^2/2$ and amplitude

$$\widetilde{g}^{(N)}(t, s, x, y) = e^{\frac{i}{\hbar}R^{(N)}(t,s,\sqrt{\hbar(t-s)}x,\sqrt{\hbar(t-s)}y)}g^{(N)}(t, s, \sqrt{\hbar(t-s)}x, \sqrt{\hbar(t-s)}y).$$

Using Proposition 7.3.3, formula (7.15) and the Banach algebra property of $M^{\infty,1}$ we have

$$\left\|e^{\frac{i}{\hbar}R^{(N)}(t,s,\cdot,\cdot)}g_N(t, s, \cdot, \cdot)\right\|_{M^{\infty,1}} \leq C(T)(t-s)^N$$

for $0 < t - s \leq T\hbar$. Again by the dilation properties in Proposition 3.2.4 we obtain

$$\left\| \tilde{g}^{(N)}(t, s, \cdot, \cdot) \right\|_{M^{\infty,1}} \leq C(T)(t - s)^N$$

for $0 < t - s \leq T\hbar$ and for a new constant $C(T) > 0$. Therefore, by Lemma 4.4.10, $G^{(N)}(t, s)$ extends to a bounded operator on $L^2(\mathbb{R}^d)$ with

$$\left\| G^{(N)} f \right\|_{L^2} \leq C \left\| \tilde{g}^{(N)} \right\|_{M^{\infty,1}} \|f\|_{L^2} \leq C(T)\,(t - s)^N\, \|f\|_{L^2}, \qquad (7.17)$$

always for $0 < t - s \leq T\hbar$.

Now, the propagator $U(t, s)$ clearly satisfies the equation

$$(i\hbar\partial_t - H)\, U(t, s)f = 0$$

for all $f \in L^2(\mathbb{R}^d)$, where $H = -\left(\hbar^2/2\right)\Delta + V$ is the Hamiltonian operator, with V as in Assumption ($\tilde{A}$). On the other hand

$$(i\hbar\partial_t - H)\, \tilde{E}^{(N)}(t, s)f = G^{(N)}\,(t, s)\, f,$$

which can be rephrased in integral form by means of Duhamel's principle as

$$\tilde{E}^{(N)}\,(t, s)\, f = U(t, s)f - \frac{i}{\hbar} \int_s^t U(t, \tau)G^{(N)}(\tau, s)f\, d\tau.$$

Therefore, given $f \in L^2\left(\mathbb{R}^d\right)$, by (7.17) we have

$$\left\| U\,(t, s)\, f - \tilde{E}^{(N)}\,(t, s)\, f \right\|_{L^2}$$

$$= \left\| \hbar^{-1} \int_s^t U(t, \tau)G^{(N)}(\tau, s)f\, d\tau \right\|_{L^2}$$

$$\leq \hbar^{-1} \int_s^t \|U(t, \tau)\|_{L^2 \to L^2} \left\| G^{(N)}(\tau, s)f \right\|_{L^2} d\tau$$

$$\leq C(T)\hbar^{-1} \int_s^t \|f\|_{L^2}\,(t - s)^N d\tau$$

$$\leq C'(T)\hbar^{-1}(t - s)^{N+1}\, \|f\|_{L^2},$$

for $0 < t - s \leq T\hbar$.   □

We are now in the position to state a convergence result for the approximate propagator $\tilde{E}^{(N)}(\Omega, t, s)$ (cf. [182, Theorem 1]).

Theorem 7.4.4 *Let V satisfy Assumption* $(\tilde{A})$ *above. For any* $T > 0$ *there exists a constant* $C = C(T) > 0$ *such that, for* $0 < t - s \leq T\hbar$, $0 < \hbar \leq 1$, *and any sufficiently fine subdivision* Ω *of the interval* $[s, t]$, *we have*

$$\|\widetilde{E}^{(N)}(\Omega, t, s) - U(t, s)\|_{L^2 \to L^2} \leq C\omega(\Omega)^N.$$

Remark 7.4.5 It comes not as a surprise that the semiclassical approximation power of Birkhoff-Maslov parametrices (1.18) is lost if the latter are replaced with rougher parametrices as those in (7.1)—which however achieve an accelerated rate of convergence with respect to time. A cursory glance at the estimates for the operators $\widetilde{E}^{(N)}$ derived in Sect. 7.4 below reveals that negative powers of $\hbar$ are involved, making them completely unfit for semiclassical arguments. Nevertheless, it is worth emphasizing that all the estimates obtained in this setting are uniform in $\hbar$ as soon as time is measured in units of $\hbar$, which is a particularly interesting feature.

Proof of Theorem 7.4.4 The claim follows at once from Theorem 7.4.3 and the abstract result in Theorem 7.2.1 applied with T replaced by $T\hbar$, $C_0 = 1$, $C_1 = C\hbar^{-1}$, where C is the constant appearing in (7.16), and using $t - s \leq T\hbar$.

Chapter 8
Convergence in $\mathcal{L}(L^2)$ for Potentials in Kato-Sobolev Spaces

8.1 Summary

In this chapter we continue the investigation on the convergence in $\mathcal{L}(L^2(\mathbb{R}^d))$ of time-slicing approximate propagators for Feynman path integrals. Specifically, we focus on a class of potentials with low regularity that still provide convergence results in L^2 for Fujiwara's parametrices as in Theorem 1.3.1 (i). The relevant function spaces are given by the Kato-Sobolev spaces $H^r_{\mathrm{ul}}(\mathbb{R}^d)$, introduced in Sect. 3.6.3. Recall that

$$\|f\|_{H^r_{\mathrm{ul}}} = \sup_{y \in \mathbb{R}^d} \|g(\cdot - y)f\|_{H^r} < \infty,$$

where $g \in C_c^\infty(\mathbb{R}^d) \setminus \{0\}$ and therefore

$$H^r_{\mathrm{ul}}(\mathbb{R}^d) = W(H^r, L^\infty)(\mathbb{R}^d) = W(\mathcal{F}L^2_r, L^\infty)(\mathbb{R}^d) = W^{2,\infty}_{r,0}(\mathbb{R}^d) = \mathcal{F}M^{2,\infty}_{r,0}(\mathbb{R}^d);$$

see Sect. 3.3 for these characterizations in terms of Wiener amalgam spaces. In particular, functions in $H^r_{\mathrm{ul}}(\mathbb{R}^d)$ for $r > d/2$ are bounded and continuous but need not decay at infinity.

To be precise, we still consider the Schrödinger equation

$$i\hbar\partial_t \psi = -\frac{1}{2}\hbar^2 \Delta\psi + V(t,x)\psi$$

as in the previous chapter, but now V (real valued) is allowed to have at most quadratic growth.

Assumption (A$'$) $V(t, x)$ *is a real-valued function such that* $\partial_x^\alpha V \in C(\mathbb{R} \times \mathbb{R}^d)$ *for* $|\alpha| \leq 2$ *and moreover*

$$\|\partial_x^\alpha V(t, \cdot)\|_{H_{ul}^{d+1}} \leq C \quad \text{for } |\alpha| = 2, \ t \in \mathbb{R}. \tag{8.1}$$

Derivatives are understood in the distribution sense. Observe that $\partial_x^\alpha V \in L^\infty(\mathbb{R} \times \mathbb{R}^d)$ for $|\alpha| = 2$, in view of the embedding $H^{d+1}(\mathbb{R}^d) \subset L^\infty(\mathbb{R}^d)$.

We show that, under this assumption, the time-slicing approach via the parametrices $E^{(0)}(\Omega, t, s)$ elaborated by Fujiwara (see Sect. 1.3.1) is still valid and a convergence result in $\mathcal{L}(L^2(\mathbb{R}^d))$ identical to Theorem 1.3.1 (i) ($N = 0$) can be proved—see Theorem 8.4.3 below for a precise statement.

Let us briefly discuss the choice of the Sobolev exponent $d+1$ appearing in (8.1). It seems that all the known results about the continuity in $L^2(\mathbb{R}^d)$ of oscillatory integral operators with non-smooth phase (and amplitude) similar to those involved here, namely

$$E^{(0)}(t, s)f(x) = \frac{1}{(2\pi i(t-s)\hbar)^{d/2}} \int_{\mathbb{R}^d} e^{\frac{i}{\hbar}S(t,s,x,y)} f(y)\,dy, \tag{8.2}$$

require that the second derivatives of the phase have, roughly speaking, at least r additional derivatives both with respect to x and y, for some $r > d/2$ (the derivatives are understood in the L^2-Sobolev scale; see Sect. 4.4.3 [27, 220] and the references therein). To offer a heuristic explanation of this occurrence we call into question some known counterexamples on the continuity in L^2 of pseudodifferential operators (see e.g. [27, Remark 2]), since the second derivatives of the phase and the amplitude often play the same role in the estimates (cf. (8.28) below). In this respect, we will prove that under the Assumption (A$'$) the classical action $S(t, s, x, y)$ has second space derivatives in $H_{ul}^{d+1}(\mathbb{R}^{2d})$, hence possessing $d/2 + 1/2$ additional derivatives with respect to both x and y. One could probably refine the result by means of fractional Kato-Sobolev spaces $H_{ul}^r(\mathbb{R}^d)$ for some $r > d$, but we prefer not to engage in further technicalities.

Even if the proof ultimately follows the same strategy used in the smooth scenario [108, 109], as a preliminary step we are required to perform a refined short-time analysis of both the Hamiltonian flow and the classical action at this low level of regularity. Sections 8.2 and 8.3 are devoted to these technical issue, using several results already collected in Sect. 3.6.3.

A careful analysis of the parametrices is carried out in Sect. 8.4, together with the proof of the main convergence result. We stress that, under additional space regularity for the potential, convergence results in the same spirit can be established for higher-order approximate propagators, with positive powers of $\hbar$ appearing in the estimates, as in Theorem 1.3.1 (i) ($N \geq 1$), hence making this analysis also suitable for semiclassical approximation purposes. The details on this problem are to be found in Sect. 8.5 below.

8.2 Sobolev Regularity of the Hamiltonian Flow

In this section we assume that the following condition on the potential $V(t, x)$ is satisfied.

Let $\kappa \in \mathbb{N}, \kappa \geq d + 1$.

Assumption (B') $V(t, x)$ *is a real-valued function such that* $\partial_x^\alpha V \in C(\mathbb{R} \times \mathbb{R}^d)$ *for* $|\alpha| \leq 2$ *and*

$$\|\partial_x^\alpha V(t, \cdot)\|_{H_{ul}^\kappa} \leq C \quad \text{for } |\alpha| = 2, \ t \in \mathbb{R}.$$

In particular, we will take $\kappa = d + 1$ in Theorem 8.4.3 and higher values of κ in Theorem 8.5.2 below.

Denote by $(x(t, s, y, \eta), \xi(t, s, y, \eta))$, $s, t \in \mathbb{R}$, $y, \eta \in \mathbb{R}^d$, the solution of the Hamiltonian system

$$\dot{x} = \xi, \quad \dot{\xi} = -\nabla_x V(t, x) \tag{8.3}$$

with initial condition at time $t = s$ given by $x(s, s, y, \eta) = y, \xi(s, s, y, \eta) = \eta$.

Note that $\partial_x^\alpha V \in C(\mathbb{R} \times \mathbb{R}^d)$ for $|\alpha| = 1$ and $\partial_x^\alpha V \in C_b(\mathbb{R} \times \mathbb{R}^d)$ for $|\alpha| = 2$, so that the solutions $x(t, s, y, \eta)$ and $\xi(t, s, y, \eta)$ above are defined for every $t \in \mathbb{R}$ and are C^1 functions of (t, y, η). Moreover, the flow

$$(y, \eta) \mapsto (x(t, s, y, \eta), \xi(t, s, y, \eta))$$

defines a C^1 canonical transformation $\mathbb{R}^d \times \mathbb{R}^d \to \mathbb{R}^d \times \mathbb{R}^d$.

Concerning quantitative information, by recasting the previous system in integral form and by Gronwall's inequality (2.4) it follows that for any $T > 0$ there exists a constant $C = C(T) > 0$ such that

$$\left\|\frac{\partial x}{\partial y}(t, s)\right\|_{L^\infty(\mathbb{R}_y^d \times \mathbb{R}_\eta^d)} + \left\|\frac{\partial x}{\partial \eta}(t, s)\right\|_{L^\infty(\mathbb{R}_y^d \times \mathbb{R}_\eta^d)}$$

$$+ \left\|\frac{\partial \xi}{\partial y}(t, s)\right\|_{L^\infty(\mathbb{R}_y^d \times \mathbb{R}_\eta^d)} + \left\|\frac{\partial \xi}{\partial \eta}(t, s)\right\|_{L^\infty(\mathbb{R}_y^d \times \mathbb{R}_\eta^d)} \leq C, \tag{8.4}$$

if $|t - s| \leq T$. In particular, the map $(x(t, s), \xi(t, s))$ is globally Lipschitz.

In fact, we can prove the following result, which is finer when $|t - s|$ is small.

Lemma 8.2.1 *For every* $T > 0$ *there exists a constant* $C = C(T) > 0$ *such that, for* $|t - s| \leq T$,

$$\left\|\frac{\partial x}{\partial y}(t, s) - I\right\|_{L^\infty(\mathbb{R}_y^d \times \mathbb{R}_\eta^d)} \leq C|t - s|^2, \quad \left\|\frac{\partial x}{\partial \eta}(t, s)\right\|_{L^\infty(\mathbb{R}_y^d \times \mathbb{R}_\eta^d)} \leq C|t - s| \tag{8.5}$$

$$\left\|\frac{\partial \xi}{\partial y}(t, s)\right\|_{L^\infty(\mathbb{R}_y^d \times \mathbb{R}_\eta^d)} \leq C|t - s|, \quad \left\|\frac{\partial \xi}{\partial \eta}(t, s) - I\right\|_{L^\infty(\mathbb{R}_y^d \times \mathbb{R}_\eta^d)} \leq C|t - s|^2.$$

(8.6)

Proof From (8.3) we have

$$\begin{cases} \dfrac{\partial x}{\partial u}(t, s) = \dfrac{\partial y}{\partial u} + \displaystyle\int_s^t \dfrac{\partial \xi}{\partial u}(\tau, s) \, d\tau \\[4mm] \dfrac{\partial \xi}{\partial u}(t, s) = \dfrac{\partial \eta}{\partial u} - \displaystyle\int_s^t \dfrac{\partial^2}{\partial x \partial x} V(\tau, x(\tau, s)) \dfrac{\partial x}{\partial u}(\tau, s) \, d\tau, \end{cases}$$

where u stands for y_j or η_j, $j = 1, \ldots, d$. Hence we obtain

$$\frac{\partial x}{\partial u}(t, s) - \frac{\partial y}{\partial u} = \int_s^t \frac{\partial \xi}{\partial u}(\tau, s) \, d\tau$$

and

$$\frac{\partial \xi}{\partial u}(t, s) - \frac{\partial \eta}{\partial u} + \int_s^t \frac{\partial^2}{\partial x \partial x} V(\tau, x(\tau, s)) \frac{\partial y}{\partial u} \, d\tau$$

$$= -\int_s^t \int_s^\tau \frac{\partial^2}{\partial x \partial x} V(\tau, x(\tau, s)) \frac{\partial \xi}{\partial u}(\sigma, s) \, d\sigma.$$

Then, using (8.4) we obtain at once the desired estimates. □

We prove now that the flow $(x(t, s, y, \eta), \xi(t, s, y, \eta))$ enjoys the same space regularity as $\nabla_x V$, namely that the first space derivatives of $x(t, s, y, \eta), \xi(t, s, y, \eta)$ belong to $H_{\mathrm{ul}}^\kappa(\mathbb{R}^{2d})$. In fact, we have the following precise asymptotic estimates of their Sobolev norm as $|t - s| \to 0$, which will play a key role below. Below we denote by $D_x^2 V$ the Hessian of V with respect to x.

Proposition 8.2.2 *Let $T > 0$. There exists a constant $C > 0$ depending only on T and upper bounds for the norm of the Hessian $D_x^2 V$ in $L^\infty(\mathbb{R}; H_{\mathrm{ul}}^\kappa(\mathbb{R}^d; \mathbb{R}^{d \times d}))$ such that, for $2 \leq |\alpha| + |\beta| \leq \kappa + 1$, $|t - s| \leq T$,*

$$\|\partial_y^\alpha \partial_\eta^\beta x(t, s)\|_{L^2(B)} \leq C|t - s|^{|\beta| + 2}, \quad \|\partial_y^\alpha \partial_\eta^\beta \xi(t, s)\|_{L^2(B)} \leq C|t - s|^{|\beta| + 1}$$

(8.7)

for every open ball $B \subset \mathbb{R}_y^d \times \mathbb{R}_\eta^d$ of radius 1.

Proof For the sake of clarity, we find it convenient to isolate the main steps of the proof.

Step 1 *Regularization.*

We commence by noticing that it suffices to prove the desired estimates if $V(t, x)$ is additionally assumed to be smooth with respect to x for every $t \in \mathbb{R}$. While the arguments below are likely well known to a reader familiar with the analysis of PDEs, we prefer to provide a detailed proof for completeness.

First, assume that the result holds for smooth potentials. Consider then smooth regularizations

$$V_\epsilon(t, \cdot) = V(t, \cdot) * \rho_\epsilon, \quad 0 < \epsilon \leq 1,$$

where $\rho_\epsilon(x) = \epsilon^{-d}\rho(\epsilon^{-1}x)$ is a standard mollifier in $\mathbb{R}^d$, i.e. $\rho \in C_c^\infty(\mathbb{R}^d)$, $\rho \geq 0$, $\int_{\mathbb{R}^d} \rho(x) = 1$. Indeed, we have $\partial_x^\alpha V_\epsilon \in C(\mathbb{R} \times \mathbb{R}^d)$ for every $\alpha \in \mathbb{N}^d$.

We have $\|D_x^2 V_\epsilon(t, \cdot)\|_{L^\infty} \lesssim \|D_x^2 V_\epsilon(t, \cdot)\|_{H_{ul}^\kappa} \lesssim \|D_x^2 V(t, \cdot)\|_{H_{ul}^\kappa} \leq C$ for a constant C independent of ϵ, so that the corresponding solution $(x_\epsilon(t, s), \xi_\epsilon(t, s))$ is defined for every $t \in \mathbb{R}$ and satisfies the estimates in (8.7) with a constant C independent of ϵ.

On the other hand, it is not difficult to verify the following facts—in the given order.

To simplify notation, set $X = (x, \xi)$, $\mathbf{b}(t, X) = (\xi, -\nabla_x V(t, x))$, $X_\epsilon = (x_\epsilon, \xi_\epsilon)$, $\mathbf{b}_\epsilon(t, X) = (\xi, -\nabla_x V_\epsilon(t, x))$, $Y = (y, \eta)$.

(i) *There exists a constant $C > 0$ such that*

$$|\nabla_x V_\epsilon(t, x) - \nabla_x V(t, 0)| \leq C(1 + |x|)$$

for every $t \in \mathbb{R}$ and $x \in \mathbb{R}^d$, $0 < \epsilon \leq 1$.

Indeed, for every $t \in \mathbb{R}$ we can perform a Taylor expansion of $\nabla_x V(t, \cdot)$ and obtain

$$\nabla_x V(t, x) - \nabla_x V(t, 0) = \mathbf{c}(t, x),$$

where $|\mathbf{c}(t, x)| \leq C|x|$ for every $t \in \mathbb{R}$ and every $x \in \mathbb{R}^d$, because $\nabla_x V(t, \cdot) \in C^1(\mathbb{R}^d)$ and $\partial_x^\alpha V \in L^\infty(\mathbb{R} \times \mathbb{R}^d)$ for $|\alpha| = 2$. Since

$$\nabla_x V_\epsilon(t, \cdot) - \nabla_x V(t, 0) = \mathbf{c}(t, \cdot) * \rho_\epsilon,$$

we easily obtain the desired conclusion.

(ii) *The solutions $X_\epsilon(t, s, Y)$ for fixed $s \in \mathbb{R}$ are bounded on the compact subsets of $\mathbb{R} \times \mathbb{R}^{2d}$, uniformly with respect to ϵ.*

In fact, since $X_\epsilon(t, s, Y) = (x_\epsilon(t, s, Y), \xi_\epsilon(t, s, Y))$ is the flow of $\mathbf{b}_\epsilon(t, X)$ we have

$$X_\epsilon(t, s, Y) = Y + \int_s^t (0, -\nabla_x V(t, 0)) d\tau$$

$$+ \int_s^t (\xi_\epsilon(\tau, s, Y), -\nabla_x V_\epsilon(\tau, x_\epsilon(\tau, s, Y)) + \nabla_x V(t, 0)) d\tau.$$

The estimates in the previous item and Gronwall's inequality (Theorem 2.5.2) give the desired conclusion.

(iii) *For every $t \in \mathbb{R}$, $\mathbf{b}_\epsilon(t, X)$ converges to $\mathbf{b}(t, X)$ as $\epsilon \to 0$, uniformly on the compact subsets of $\mathbb{R}^{2d}$.*

(iv) *The difference $\mathbf{b}_\epsilon(t, X) - \mathbf{b}(t, X)$ is bounded on the compact subsets of $\mathbb{R} \times \mathbb{R}^{2d}$, uniformly with respect to ϵ.*

Indeed, with the notation in item (i), we have $\mathbf{c} \in C(\mathbb{R} \times \mathbb{R}^d; \mathbb{R}^d)$. On the other hand

$$\nabla_x V_\epsilon(t, \cdot) - \nabla_x V(t, \cdot) = \mathbf{c}(t, \cdot) * \rho_\epsilon - \mathbf{c}(t, \cdot)$$

and hence the claim follows.

(v) *For fixed $s \in \mathbb{R}$, $X_\epsilon(t, s, Y)$ converges to $X(t, s, Y)$ as $\epsilon \to 0$, uniformly on the compact subsets of $\mathbb{R} \times \mathbb{R}^d$.*

In fact it turns out that for fixed $T, R > 0$, by the point (ii) there exists a ball $B' \subset \mathbb{R}^{2d}$ where the functions $X_\epsilon(t, s, Y)$ take values for $|t| \leq T$, $|Y| \leq R$, and we have[1]

$$|X_\epsilon(t, s, Y) - X(t, s, Y)| \leq \int_s^t \|\mathbf{b}_\epsilon(\tau, \cdot) - \mathbf{b}(\tau, \cdot)\|_{L^\infty(B')} d\tau$$

$$+ \int_s^t \|\nabla_X \mathbf{b}(\tau, \cdot)\|_{L^\infty(\mathbb{R}^{2d})} |X_\epsilon(\tau, s, Y) - X(\tau, s, Y)| d\tau.$$

The first integral vanishes due to the dominated convergence theorem (see the items (iii) and (iv)) and one concludes by means of Gronwall's inequality (2.4).

Therefore, for every open ball $B \subset \mathbb{R}^{2d}$ the functions $X_\epsilon(t, s, \cdot) = (x_\epsilon(t, s), \xi_\epsilon(t, s))$ converge in $\mathcal{D}'(B)$ to $X(t, s) = (x(t, s), \xi(t, s))$ as $\epsilon \to 0$, and then the desired estimates (8.7) hold for $(x(t, s), \xi(t, s))$. In fact, as a consequence of the Riesz representation theorem and the density of $C_c^\infty(B)$ in $L^2(B)$, if u_n is a

---

[1] The L^∞ norm of a vector-valued function f is understood here as the L^∞ norm of the Euclidean norm $|f(x)|$.

bounded sequence in $L^2(B)$, say $\|u_n\|_{L^2(B)} \leq C$, converging to a distribution u in $\mathcal{D}'(B)$, it turns out $u \in L^2(B)$ and $\|u\|_{L^2(B)} \leq C$.

Step 2. *A priori estimates.*

Let us now prove the formulas (8.7) as priori estimates. Namely, we assume that $\partial_x^\alpha V \in C(\mathbb{R} \times \mathbb{R}^d)$ for all $\alpha \in \mathbb{N}^d$, which implies that $x(t, s, y, \eta)$ and $\xi(t, s, y, \eta)$ are smooth with respect to $y, \eta \in \mathbb{R}^d$. In order to properly argue by induction we need to state the inductive hypothesis in a slightly stronger form. Precisely, for

$$2 \leq |\alpha| + |\beta| \leq \kappa + 1, \quad |t - s| \leq T, \quad 1/p = (|\alpha| + |\beta| - 1)/(2\kappa),$$

we have

$$\|\partial_y^\alpha \partial_\eta^\beta x(t, s)\|_{L^p(B)} \leq C|t - s|^{|\beta|+2}, \tag{8.8}$$

$$\|\partial_y^\alpha \partial_\eta^\beta \xi(t, s)\|_{L^p(B)} \leq C|t - s|^{|\beta|+1} \tag{8.9}$$

for a constant $C > 0$ independent of B, as in the statement. Therefore we will prove these estimates by induction on $|\alpha| + |\beta|$.

Observe that, for $2 \leq |\alpha| + |\beta| \leq \kappa + 1$,

$$\partial_y^\alpha \partial_\eta^\beta x(t, s, y, \eta) = \int_s^t \partial_y^\alpha \partial_\eta^\beta \xi(\tau, s, y, \eta) \, d\tau \tag{8.10}$$

and

$$\partial_y^\alpha \partial_\eta^\beta \xi(t, s, y, \eta) = -\int_s^t \sum_{j=1}^d (\partial_{x_j} \nabla_x V)(\tau, x(\tau, s)) \partial_y^\alpha \partial_\eta^\beta x_j(\tau, s, y, \eta) \, d\tau$$

$$- \int_s^t b_{\alpha,\beta}(\tau, s, y, \eta) \, d\tau, \tag{8.11}$$

where $b_{\alpha,\beta}(\tau, s, y, \eta)$ is a linear combination of terms of the form

$$(\partial_x^\sigma \nabla_x V)(\tau, x(\tau, s)) \left(\partial_y^{\nu_1} \partial_\eta^{\mu_1} x_{j_1}(\tau, s) \right) \dots \left(\partial_y^{\nu_{|\sigma|}} \partial_\eta^{\mu_{|\sigma|}} x_{j_{|\sigma|}}(\tau, s) \right)$$

and $j_1, \dots, j_{|\sigma|} \in \{1, \dots, d\}$, $\nu_1 + \dots + \nu_{|\sigma|} = \alpha$, $\mu_1 + \dots + \mu_{|\sigma|} = \beta$, $|\nu_j| + |\mu_j| \geq 1$ for $j = 1, \dots, |\sigma|$ and $2 \leq |\sigma| \leq |\alpha| + |\beta|$.

We can estimate the norm in $L^p(B)$ of this expression, with

$$\frac{1}{p} = \frac{|\alpha| + |\beta| - 1}{2\kappa},$$

by Hölder's inequality as

$$\|(\partial_x^\sigma \nabla_x V)(\tau, x(\tau, s))\|_{L^{p_0}(B)} \|\partial_y^{\nu_1} \partial_\eta^{\mu_1} x\|_{L^{p_1}(B)} \cdots \|\partial_y^{\nu_{|\sigma|}} \partial_\eta^{\mu_{|\sigma|}} x\|_{L^{p_{|\sigma|}}(B)}, \qquad (8.12)$$

where

$$\frac{1}{p_0} = \frac{|\sigma| - 1}{2\kappa}, \qquad \frac{1}{p_j} = \frac{|\nu_j| + |\mu_j| - 1}{2\kappa}, \quad j = 1, \ldots, |\sigma|.$$

Notice that

$$\sum_{j=0}^{|\sigma|} \frac{1}{p_j} = \frac{|\sigma| - 1 + |\alpha| + |\beta| - |\sigma|}{2\kappa} = \frac{1}{p}.$$

Since the map $(x(t, s), \xi(t, s))$ is a canonical transformation, hence measure preserving, the first factor in (8.12) can be written as

$$\|(\partial_x^\sigma \nabla_x V)(\tau, x(\tau, s))\|_{L^{p_0}(B)} = \|(\partial_x^\sigma \nabla_x V)(\tau, \cdot)\|_{L^{p_0}(\tilde{B})},$$

where $\tilde{B} \subset \mathbb{R}_x^d \times \mathbb{R}_\xi^d$ is the image of B by the flow $(x(t, s), \xi(t, s))$—here we consider $V(\tau, x)$ as a function of τ, x, ξ, constant with respect to ξ. Now, using (8.4), we see that $\tilde{B}$ can be covered by $N = N(T, d)$ boxes $B' \times B''$, with $B', B'' \subset \mathbb{R}^d$ balls of radius 1, for $|t - s| \le T$, so that we can continue the estimate, by the Gagliardo-Nirenberg-Sobolev inequality (2.2) (using $2 \le |\sigma| \le \kappa + 1$) as

$$\|(\partial_x^\sigma \nabla_x V)(\tau, x(\tau, s))\|_{L^{p_0}(B)} \le C \sup_{B''} \|\partial_x^\sigma \nabla_x V(\tau, \cdot)\|_{L^{p_0}(B'')}$$

$$\le C' \sup_{B''} \|D_x^2 V\|_{L^\infty(\mathbb{R} \times \mathbb{R}^d)}^{1-(|\sigma|-1)/\kappa} \|D_x^2 V\|_{L^\infty(\mathbb{R}; H^\kappa(B''))}^{(|\sigma|-1)/\kappa}$$

$$\le C' \|D_x^2 V\|_{L^\infty(\mathbb{R} \times \mathbb{R}^d)}^{1-(|\sigma|-1)/\kappa} \|D_x^2 V\|_{L^\infty(\mathbb{R}; H_{ul}^\kappa)}^{(|\sigma|-1)/\kappa}.$$

The latter quantity is finite, since by assumption we have

$$\|D_x^2 V\|_{L^\infty(\mathbb{R} \times \mathbb{R}^d)} \le C \|D_x^2 V\|_{L^\infty(\mathbb{R}; H_{ul}^\kappa)} < \infty.$$

In order to handle the other factors in (8.12) it is useful to observe that

$$\|\partial_y^{\nu_j} \partial_\eta^{\mu_j} x(\tau, s)\|_{L^{p_j}(B)} \le C |\tau - s|^{|\mu_j|}$$

for a constant $C = C(T)$ if $|\tau - s| \leq T$. This holds for $|\nu_j| + |\mu_j| = 1$ by (8.5) and if $|\nu_j| + |\mu_j| \geq 2$ by the inductive hypothesis (because $|\nu_j| + |\mu_j| \leq |\alpha| + |\beta| - 1$, therefore in the latter case we have $|\alpha| + |\beta| \geq 3$).

In conclusion, since $\mu_1 + \ldots + \mu_{|\sigma|} = \beta$ we have

$$\|b_{\alpha,\beta}(\tau, s)\|_{L^p(B)} \leq C|\tau - s|^{|\beta|}$$

and we can estimate in (8.11)

$$\|\partial_y^\alpha \partial_\eta^\beta \xi(t, s)\|_{L^p(B)} \leq C \int_s^t \|\partial_y^\alpha \partial_\eta^\beta x(\tau, s)\|_{L^p(B)} \, d\tau + C|t - s|^{|\beta|+1}.$$

Combining the latter estimate with (8.10) and Gronwall's inequality (2.4) gives (8.9) and also (8.8).   □

Consider now the map

$$(y, \zeta) \mapsto (\tilde{x}(t, s, y, \zeta), \tilde{\xi}(t, s, y, \zeta)) := (x(t, s, y, \zeta/(t-s)), (t-s)\xi(t, s, y, \zeta/(t-s)),$$

with $s, t \in \mathbb{R}$ $s \neq t$, $y, \zeta \in \mathbb{R}^d$. Observe that it is a C^1 canonical transformation.
We have the following result.

Proposition 8.2.3 *We have, for $j, k = 1, \ldots, d$,*

$$\frac{\partial \tilde{x}_j}{\partial \zeta_k} = \delta_{j,k} - (t - s)^2 a_{jk}(t, s, y, \zeta), \tag{8.13}$$

$$\frac{\partial \tilde{\xi}_j}{\partial \zeta_k} = \delta_{j,k} - (t - s)^2 b_{jk}(t, s, y, \zeta), \tag{8.14}$$

$$\frac{\partial \tilde{x}_j}{\partial y_k} = \delta_{j,k} - (t - s)^2 c_{jk}(t, s, y, \zeta), \tag{8.15}$$

$$\frac{\partial \tilde{\xi}_j}{\partial y_k} = (t - s)^2 c'_{jk}(t, s, y, \zeta),$$

where $a_{jk}(t, s), b_{jk}(t, s), c_{jk}(t, s), c'_{jk}(t, s)$ are functions in a bounded subset of $H_{ul}^\kappa(\mathbb{R}^{2d})$, if $0 < |t - s| \leq T$, for every fixed $T > 0$.

Proof We discuss the details of the proof only for formula (8.14), since the other ones can be proved by means of pretty similar arguments.

First of all, we observe that (8.14) defines the functions $b_{jk}(t, s)$, since $s \neq t$. Now, if $B \subset \mathbb{R}^d_y \times \mathbb{R}^d_\zeta$ is any open ball of radius 1 we have clearly that, by the second formula in (8.6) with $\eta = \zeta/(t - s)$,

$$\|b_{jk}(t, s)\|_{L^2(B)} = (t - s)^{-2}\|\frac{\partial \xi_j}{\partial \eta_k}(t, s, y, \zeta/(t - s)) - \delta_{jk}\|_{L^2(B)}$$

$$\leq C(t - s)^{-2}\|\frac{\partial \xi_j}{\partial \eta_k}(t, s, y, \zeta/(t - s)) - \delta_{jk}\|_{L^\infty(B)} \leq C'$$

for $0 < |t - s| \leq T$ and some $C' = C'(T) > 0$ independent of B.

We now estimate the derivatives of b_{jk}. For $1 \leq |\alpha| + |\beta| \leq \kappa$ we have

$$\partial_y^\alpha \partial_\zeta^\beta b_{jk}(t, s, y, \zeta) = -(t - s)^{-2-|\beta|}\left(\partial_y^\alpha \partial_\eta^\beta \frac{\partial \xi_j}{\partial \eta_k}\right)(t, s, y, \zeta/(t - s))$$

and therefore

$$\|\partial_y^\alpha \partial_\zeta^\beta b_{jk}(t, s)\|_{L^2(B)} \leq |t - s|^{-2-|\beta|+d/2}\left\|\partial_y^\alpha \partial_\eta^\beta \frac{\partial \xi_j}{\partial \eta_k}(t, s)\right\|_{L^2(\tilde{B})},$$

where

$$\tilde{B} = \{(y, \eta) \in \mathbb{R}^d \times \mathbb{R}^d : (y, (t - s)\eta) \in B\}.$$

Observe that $\tilde{B}$ can be covered by $N = O(|t - s|^{-d})$ balls $B' \subset \mathbb{R}^d_y \times \mathbb{R}^d_\eta$ of radius 1, and therefore

$$\left\|\partial_y^\alpha \partial_\eta^\beta \frac{\partial \xi_j}{\partial \eta_k}(t, s)\right\|_{L^2(\tilde{B})} \leq C|t-s|^{-d/2} \sup_{B'} \left\|\partial_y^\alpha \partial_\eta^\beta \frac{\partial \xi_j}{\partial \eta_k}(t, s)\right\|_{L^2(B')} \leq C'|t-s|^{-d/2+|\beta|+2},$$

where in the last inequality we used the second formula in (8.7).

To summarize, we have

$$\|\partial_y^\alpha \partial_\zeta^\beta b_{jk}(t, s)\|_{L^2(B)} \leq C$$

for some $C = C(T) > 0$ independent of B, for $0 < |t - s| \leq T$. $\square$

It follows from the proof, but also from the inclusion $H_{ul}^\kappa(\mathbb{R}^{2d}) \subset L^\infty(\mathbb{R}^{2d})$ (recall $\kappa \geq d + 1$), that the functions $a_{jk}(t, s), b_{jk}(t, s), c_{jk}(t, s), c'_{jk}(t, s)$ belong to a bounded subset of $L^\infty(\mathbb{R}^{2d})$ for $0 < |t - s| \leq T$.

The previous results converge to form the following regularity result.

Proposition 8.2.4 *For any fixed $T > 0$, the components of the canonical transformation $(\tilde{x}(t, s, y, \zeta), \tilde{\xi}(t, s, y, \zeta))$, have first space derivatives in a bounded subset of $H_{ul}^\kappa(\mathbb{R}^{2d})$, when $0 < |t - s| \leq T$.*

We now study the regularity of the inverse function of $\zeta \mapsto \tilde{x}(t, s, y, \zeta)$, for $|t - s|$ sufficiently small.

Proposition 8.2.5 *There exists $\delta > 0$ such that for $0 < |t - s| \leq \delta$ and $y \in \mathbb{R}^d$ the map*

$$\zeta \mapsto \tilde{x}(t, s, y, \zeta) = x(t, s, y, \zeta/(t - s))$$

is invertible $\mathbb{R}^d \to \mathbb{R}^d$ and its inverse $\zeta = \zeta(t, s, \tilde{x}, y)$ has first derivatives with respect to $\tilde{x}$ and y in a bounded subset of $H^\kappa_{ul}(\mathbb{R}^{2d})$.
  More precisely, for $j, k = 1, \ldots, d$, we have

$$\frac{\partial \zeta_j}{\partial \tilde{x}_k}(t, s, \tilde{x}, y) = \delta_{jk} - (t - s)^2 d_{jk}(t, s, \tilde{x}, y) \tag{8.16}$$

and

$$\frac{\partial \zeta_j}{\partial y_k}(t, s, \tilde{x}, y) = -\delta_{jk} - (t - s)^2 d'_{jk}(t, s, \tilde{x}, y) \tag{8.17}$$

where $d_{jk}(t, s, \tilde{x}, y)$ and $d'_{jk}(t, s, \tilde{x}, y)$ belong to a bounded subset of $H^\kappa_{ul}(\mathbb{R}^d_{\tilde{x}} \times \mathbb{R}^d_y)$ for $0 < |t - s| \leq \delta$.

Proof Since $H^\kappa_{ul}(\mathbb{R}^{2d})$ is a Banach algebra ($\kappa \geq d + 1$), by (8.13) we have

$$\det \frac{\partial \tilde{x}}{\partial \zeta}(t, s, y, \zeta) = 1 - (t - s)^2 a(t, s, y, \zeta) \tag{8.18}$$

for some function $a(t, s, y, \zeta)$ in a bounded subset of $H^\kappa_{ul}(\mathbb{R}^{2d}) \subset L^\infty(\mathbb{R}^{2d})$, for $0 < |t - s| \leq T$, for any $T > 0$. Choose $\delta > 0$ such that

$$(t - s)^2 \|a(t, s, \cdot, \cdot)\|_{L^\infty(\mathbb{R}^{2d})} \leq 1/2 \tag{8.19}$$

for $0 < |t - s| \leq \delta$.
  The invertibility of the C^1 map $\zeta \mapsto \tilde{x}(t, s, y, \zeta)$, for $0 < |t - s| \leq \delta$, then follows from Hadamard's global inversion theorem [41, Theorem 2, page 93]. Moreover, for $0 < |t - s| \leq \delta$ the map $(y, \zeta) \mapsto (\tilde{x}(t, s, y, \zeta), y)$ is globally bi-Lipschitz, uniformly with respect to t, s, with first derivatives in a bounded subset of $H^\kappa_{ul}(\mathbb{R}^{2d})$ (by (8.13)). By Proposition 3.6.7 we see that the same holds for the inverse map $(\tilde{x}, y) \mapsto (y, \zeta(t, s, \tilde{x}, y))$.
  In order to prove (8.16) and (8.17) we use the formulas

$$\frac{\partial \zeta}{\partial \tilde{x}}(t, s, \tilde{x}, y) = \left[\frac{\partial \tilde{x}}{\partial \zeta}(t, s, y, \zeta) \right]^{-1} \quad \text{with } \zeta = \zeta(t, s, \tilde{x}, y),$$

and

$$\frac{\partial \zeta}{\partial y}(t, s, \tilde{x}, y) = -\left[\frac{\partial \tilde{x}}{\partial \zeta}(t, s, y, \zeta)\right]^{-1} \frac{\partial \tilde{x}}{\partial y}(t, s, y, \zeta) \quad \text{with } \zeta = \zeta(t, s, \tilde{x}, y).$$

We claim that it suffices to prove that

$$\left[\frac{\partial \tilde{x}}{\partial \zeta}(t, s, y, \zeta)\right]^{-1} = [\delta_{jk} - (t - s)^2 d''_{jk}(t, s, y, \zeta)]_{j,k=1,\ldots,d} \tag{8.20}$$

with $d''_{jk}(t, s, y, \zeta)$ belonging to a bounded subset of $H^{\kappa}_{\mathrm{ul}}(\mathbb{R}^{2d})$. Indeed, since $H^{\kappa}_{\mathrm{ul}}(\mathbb{R}^{2d})$ is a Banach algebra, by (8.20), (8.15) and the composition property in Proposition 3.6.6 (with the map $(\tilde{x}, y) \mapsto (y, \zeta(t, s, \tilde{x}, y))$ playing the role of g) we see that (8.16) and (8.17) follow.

To prove (8.20) we observe that, by (8.18) and (8.19) we have

$$\left(\det \frac{\partial \tilde{x}}{\partial \zeta}(t, s, y, \zeta)\right)^{-1} = 1 + (t - s)^2 a'(t, s, y, \zeta),$$

where

$$a'(t, s, y, \zeta) = \frac{a(t, s, y, \zeta)}{1 - (t - s)^2 a(t, s, y, \zeta)}$$

is easily proved to belong to a bounded subset of $H^{\kappa}_{\mathrm{ul}}(\mathbb{R}^{2d})$ for $0 < |t - s| \le \delta$—it is enough to combine the chain rule with arguments similar to those appearing in the proof of Proposition 3.6.6, also using that $\kappa \ge d + 1$.

Combining the latter conclusion with (8.13) yields (8.20), and the proof is complete. $\square$

8.3 Sobolev Regularity of the Classical Action

We now study the regularity of the action $S(t, s, x, y)$, $s, t \in \mathbb{R}$, $s \ne t$, $x, y \in \mathbb{R}^d$, defined as the functional action computed along the classical path γ (i.e. the path satisfying the Euler-Lagrange equations) under the boundary condition $\gamma(s) = y$, $\gamma(t) = x$, which is unique for $|t - s|$ small, say $0 < |t - s| \le \delta$ where $\delta > 0$ is the constant appearing in Proposition 8.2.5. Hence

$$S(t, s, x, y) = \int_s^t L(\gamma(\tau), \dot{\gamma}(\tau), \tau) \, d\tau$$

$$= \int_s^t \left(\frac{1}{2}|\xi(\tau, s, y, \eta(t, s, x, y))|^2 - V(\tau, x(\tau, s, y, \eta(t, s, x, y)))\right) d\tau,$$

where $L(x, v, t) = \frac{1}{2}|v|^2 - V(t, x)$ is the Lagrangian of the corresponding classical system.

First of all, we observe that $S(t, s, x, y)$, for fixed $s \in \mathbb{R}$, is a C^1 function of (t, x, y). Moreover, classical arguments (see e.g. [104, Section 2.1] and [80, 166]) imply that

$$\frac{\partial S}{\partial x_j}(t, s, x, y) = \xi_j(t, s, y, \eta(t, s, x, y)) \tag{8.21}$$

and

$$\frac{\partial S}{\partial y_j}(t, s, x, y) = -\eta_j(t, s, x, y) \tag{8.22}$$

for $j = 1, \ldots, d$ (in particular, S is C^2 with respect to x, y), and that S satisfies the Hamilton-Jacobi equation

$$\partial_t S(t, s, x, y) + \frac{1}{2}|\nabla_x S(t, s, x, y)|^2 + V(t, x) = 0. \tag{8.23}$$

Proposition 8.3.1 *For $0 < |t - s| \leq \delta$ we have*

$$S(t, s, x, y) = \frac{1}{2}\frac{|x - y|^2}{t - s} + (t - s)\omega(t, s, x, y), \tag{8.24}$$

where the functions $\partial_x^\alpha \partial_y^\beta \omega(t, s, x, y)$, for $|\alpha| + |\beta| = 2$, belong to a bounded subset of $H_{\mathrm{ul}}^\kappa(\mathbb{R}^{2d})$.

Proof Consider for example the derivatives of S with respect to x. Using the formula

$$\omega(t, s, x, y) = (t - s)^{-2}\left[(t - s)S(t, s, x, y) - \frac{1}{2}|x - y|^2\right]$$

together with (8.21), (8.22), (8.14) and (8.16) we deduce easily (cf. [108, Formula (2.12)]) that

$$\frac{\partial^2}{\partial x_j \partial x_k}\omega(t, s, x, y) = -b_{jk}(t, s, y, \zeta(t, s, x, y)) - d_{jk}(t, s, x, y)$$

$$+ (t - s)^2 \sum_{m=1}^{d} b_{jm}(t, s, y, \zeta(t, s, x, y))d_{mk}(t, s, x, y).$$

where the functions b_{jk}, d_{jk} are defined in (8.14) and (8.16). This expression belongs to a bounded subset of $H_{\mathrm{ul}}^\kappa(\mathbb{R}^{2d})$, for $0 < |t - s| \leq \delta$, by Proposi-

tions 8.2.3, 8.2.5 and Proposition 3.6.6 (with the map $(x, y) \mapsto (y, \zeta(t, s, x, y))$
playing the role of the map g).

One can similarly deal with the other second derivatives of S, using the formulas

$$\frac{\partial^2}{\partial x_j \partial y_k} \omega(t, s, x, y) = (t - s)^{-2} \left[-\frac{\partial \zeta_k}{\partial x_j}(t, s, x, y) + \delta_{jk} \right] = d_{jk}(t, s, x, y)$$

and

$$\frac{\partial^2}{\partial y_j \partial y_k} \omega(t, s, x, y) = (t - s)^{-2} \left[-\frac{\partial \zeta_j}{\partial y_k}(t, s, x, y) - \delta_{jk} \right] = d'_{jk}(t, s, x, y),$$

where the functions $d'_{jk}(t, s, x, y)$ are defined in (8.17).   □

8.4 Analysis of the Parametrices and Convergence Results

Throughout this section we assume that $V(t, s)$ satisfies the hypothesis in Assumption (A') stated in the summary of this chapter, which corresponds to Assumption (B') in Sect. 8.2 with $\kappa = d + 1$. In particular, the whole machinery of Sects. 8.2 and 8.3 applies with this value of κ.

Note that, by virtue of Proposition 8.3.1, the operator $E^{(0)}(t, s)$ defined in (8.2) satisfies the assumptions of Proposition 4.4.9, with $s = d + 1$, $\lambda = h^{-1}/|t - s|$, for $0 < |t - s| \leq \delta$, possibly for a smaller value of δ. As a result, it does extend to a bounded operator in $L^2(\mathbb{R}^d)$ which satisfies

$$\|E^{(0)}(t, s)\|_{L^2 \to L^2} \leq C \tag{8.25}$$

for $0 < |t - s| \leq \delta$.

The goal of the next result is to show that $E^{(0)}(t, s)$ converges in the strong sense to the identity operator as $t \to s$.

Proposition 8.4.1 *For every $f \in L^2(\mathbb{R}^d)$ we have*

$$\lim_{t \to s} E^{(0)}(t, s) f = f, \qquad \text{in } L^2(\mathbb{R}^d).$$

Moreover the correspondence $(s, t) \mapsto E^{(0)}(t, s)$ is strongly continuous in $L^2(\mathbb{R}^d)$ if we define $E^{(0)}(s, s) = I$.

Proof In view of (8.25) we can assume $f \in \mathcal{S}(\mathbb{R}^d)$. We approximate the action $S(t, s, x, y)$ in (8.24) by means of a family of smooth actions

$$S_\epsilon(t, s, x, y) = \frac{1}{2} \frac{|x - y|^2}{t - s} + (t - s)\omega_\epsilon(t, s, x, y),$$

with

$$\omega_\epsilon(t, s, \cdot, \cdot) = \omega(t, s, \cdot, \cdot) * \rho_\epsilon,$$

where $\rho_\epsilon(z) = \epsilon^{-2d}\rho(z/\epsilon)$, $z = (x, y) \in \mathbb{R}^{2d}$, $\rho \in C_c^\infty(\mathbb{R}^{2d})$, $\rho \geq 0$, $\int_{\mathbb{R}^{2d}} \rho(z)dz = 1$ and $0 < \epsilon < 1$. We further assume $\int_{\mathbb{R}^{2d}} z_j \rho(z)dz = 0$, $j = 1, \ldots, 2d$, so that $\partial^\alpha \mathcal{F}^{-1}(\rho)(0) = 0$ if $|\alpha| = 1$.

By means of the estimates in Proposition 8.3.1 (with $\kappa = d + 1$ as usual in this section) we infer that $\partial_x^\alpha \partial_y^\beta \omega_\epsilon(t, s, x, y)$, $|\alpha| + |\beta| = 2$, belongs to a bounded set of $C_b^\infty(\mathbb{R}^{2d})$ for $0 < |t - s| \leq \delta$, and

$$\|\omega_\epsilon(t, s, \cdot, \cdot) - \omega(t, s, \cdot, \cdot)\|_{H_{ul}^{d+1}} \to 0 \quad \text{as } \epsilon \to 0, \tag{8.26}$$

uniformly for $0 < |t-s| \leq \delta$. We leave the details to the reader, since the arguments are similar to those appearing in the proof of the subsequent Proposition 8.4.2—here one uses the fact that the symbol $(1 - \mathcal{F}^{-1}(\rho)(\epsilon\zeta))/|\zeta|^2$, $\zeta \in \mathbb{R}^{2d}$, tends to 0 in $L^\infty(\mathbb{R}^{2d})$ as $\epsilon \to 0$, since $|1 - \mathcal{F}^{-1}(\rho)(\zeta)| \lesssim |\zeta|^2$.

Now we set

$$E_\epsilon^{(0)}(t, s)f(x) := \frac{1}{(2\pi i(t - s)\hbar)^{d/2}} \int_{\mathbb{R}^d} e^{\frac{i}{\hbar} S_\epsilon(t,s,x,y)} f(y)\,dy$$

$$= \frac{1}{(2\pi i(t - s)\hbar)^{d/2}} \int_{\mathbb{R}^d} e^{\frac{i}{\hbar} S(t,s,x,y)} e^{\frac{i}{\hbar}(t-s)(\omega_\epsilon(t,s,x,y)-\omega(t,s,x,y))} f(y)\,dy.$$

One can expand the second exponential in the Banach algebra $H_{ul}^{d+1}(\mathbb{R}^{2d})$, and Proposition 4.4.9 and (8.26) imply that

$$\|E_\epsilon^{(0)}(t, s) - E^{(0)}(t, s)\|_{L^2 \to L^2} \to 0 \quad \text{as } \epsilon \to 0,$$

uniformly for $0 < |t - s| \leq \delta$.

Therefore, it is enough to prove the desired conclusion for $E_\epsilon^{(0)}(t, s)$ for fixed ϵ; the latter follows from the classical analysis in [108, Proposition 4.3].   □

We continue the study of $E^{(0)}(t, s)$. Since $S(t, s, x, y)$ satisfies the Hamilton-Jacobi equation (8.23), using Proposition 8.3.1 we see that $E^{(0)}(t, s)$ is a parametrix in the sense that

$$\left(i\hbar\partial_t + \frac{1}{2}\hbar^2\Delta - V(t, x)\right)E^{(0)}(t, s)f = G^{(0)}(t, s)f \tag{8.27}$$

with

$$G^{(0)}(t, s)f = \frac{(1/2)i\hbar(t - s)}{(2\pi i(t - s)\hbar)^{d/2}} \int_{\mathbb{R}^d} e^{\frac{i}{\hbar} S(t,s,x,y)} \Delta_x \omega(t, s, x, y)f(y)\,dy, \tag{8.28}$$

(cf. [109, Formula (1.12)]), where $\omega(t, s, x, y)$ is defined in (8.24) (differentiation under the integral sign is straightforward for $f \in \mathcal{S}(\mathbb{R}^d)$).

It follows again from Propositions 8.3.1 (with $\kappa = d + 1$) and Proposition 4.4.9 (with $s = d + 1$) that $G^0(t, s)$ is bounded on $L^2(\mathbb{R}^d)$ and satisfies the key estimate

$$\|G^{(0)}(t, s)\|_{L^2 \to L^2} \leq C\hbar |t - s| \qquad (8.29)$$

for $0 < |t - s| \leq \delta$, possibly for a smaller value of δ.

Concerning the actual propagator, we can prove its existence for potentials more general than those in Assumption (A').

Proposition 8.4.2 *Let $V(t, x)$ be a real-valued continuous function such that $\partial_x^\alpha V \in L^\infty(\mathbb{R} \times \mathbb{R}^d)$ for $|\alpha| = 2$.*

Let $s \in \mathbb{R}$. Then the Cauchy problem

$$\begin{cases} i\hbar \partial_t \psi = -\frac{1}{2}\hbar^2 \Delta \psi + V(t, x)\psi \\ \psi(s, x) = f(x) \end{cases}$$

is forward and backward globally well-posed in $L^2(\mathbb{R}^d)$. The propagator $U(t, s)$ is unitary on $L^2(\mathbb{R}^d)$.

Proof Let $\chi(\xi)$ be a smooth real-valued function in $\mathbb{R}^d$, even, supported in $|\xi| \leq 2$, with $\chi(\xi) = 1$ for $|\xi| \leq 1$. We split the potential as

$$V(t, x) = \underbrace{\chi(D_x)V(t, x)}_{V_0(t,x)} + \underbrace{(1 - \chi(D_x))V(t, x)}_{V_1(t,x)} .$$

Then V_0 and V_1 are real-valued with $\partial_x^\alpha V_0 \in C(\mathbb{R} \times \mathbb{R}^d)$ for any $\alpha \in \mathbb{N}$ and we claim that

$$\partial_x^\alpha V_0 \in L^\infty(\mathbb{R} \times \mathbb{R}^d) \quad \text{for } |\alpha| \geq 2 \qquad (8.30)$$

and

$$V_1 \in L^\infty(\mathbb{R} \times \mathbb{R}^d). \qquad (8.31)$$

Let us prove (8.30). For $|\alpha| \geq 2$, write $\alpha = \beta + \gamma$, with $|\gamma| = 2$ and

$$\partial_x^\alpha V_0(t, x) = \chi(D_x)\partial_x^\beta(\partial_x^\gamma V(t, x)).$$

We have $\partial_x^\gamma V \in L^\infty(\mathbb{R} \times \mathbb{R}^d)$ by assumption and the Fourier multiplier $\chi(D_x)\partial_x^\beta$ is bounded in $L^\infty(\mathbb{R}^d)$, because its symbol is the compactly supported smooth function $(2\pi i)^{|\beta|}\chi(\xi)\xi^\beta$.

Concerning (8.31), we write

$$V_1(t, x) = (1 - \chi(D_x))\Delta_x^{-1}\Delta_x V(t, x).$$

By assumption $\Delta_x V \in L^\infty(\mathbb{R} \times \mathbb{R}^d)$ and the Fourier multiplier $(1 - \chi(D_x))\Delta_x^{-1}$ is bounded on $L^\infty(\mathbb{R}^d)$, because its symbol $\sigma(\xi) = -(2\pi|\xi|)^{-2}(1 - \chi(\xi))$ is smooth in $\mathbb{R}^d$ and has (inverse) Fourier transform in $L^1(\mathbb{R}^d)$; indeed, repeated integrations by parts give $|x|^k \hat{\sigma}(x) \in L^\infty$ for every $k \geq d - 1$.

We emphasize that the potential $V_0(t, x)$ falls in the class considered in [108, 109, 111], and the corresponding propagator $U_0(t, s)$ was constructed in [109, Theorem 3.1] in terms of a family of unitary operators in $L^2(\mathbb{R}^d)$. The original Cauchy problem in presence of a perturbation can be recast in integral form as

$$\psi(t, \cdot) = U_0(t, s)f - \frac{i}{\hbar} \int_s^t U_0(t, \tau)V_1(\tau, \cdot)\psi(\tau, \cdot)\,d\tau.$$

The latter is a generalized Volterra integral equation. Using (8.31) it is easy to show that there is a unique solution in $C([s - T, s + T]; L^2(\mathbb{R}^d))$, for every $T > 0$. Unitarity of $U(t, s)$ follows by means of standard energy estimates. $\qquad\square$

Given any subdivision $\Omega : s = t_0 < t_1 < \ldots < t_L = t$ of the interval $[s, t]$ such that $\omega(\Omega) := \max\{t_j - t_{j-1}, j = 1, \ldots, L\} \leq \delta$, we define the operator

$$E^{(0)}(\Omega, t, s) = E^{(0)}(t, t_{L-1})E^{(0)}(t_{L-1}, t_{L-2})\ldots E^{(0)}(t_1, s).$$

Theorem 8.4.3 *Suppose that the condition in Assumption (A') holds. For every $T > 0$ there exists $C = C(T) > 0$ such that for $0 < t - s \leq T$ and any subdivision Ω of the interval $[s, t]$ with $\omega(\Omega) \leq \delta$, and $0 < \hbar \leq 1$, we have*

$$\|E^{(0)}(\Omega, t, s) - U(t, s)\|_{L^2 \to L^2} \leq C\omega(\Omega)(t - s).$$

We notice that the rate of convergence of $E^{(0)}(\Omega, t, s)$ is the same as that which appears in Theorem 1.3.1 (cf. [108, 109]) for smooth potentials.

Proof Consider the operator $R^{(0)}(t, s)$ defined by

$$R^{(0)}(t, s) = E^{(0)}(t, s) - U(t, s).$$

By Proposition 8.4.1 and (8.27) we can write

$$R^{(0)}(t, s)f = -\frac{i}{\hbar} \int_s^t U(t, \tau)G^{(0)}(\tau, s)f\,d\tau.$$

Using (8.29) we deduce that

$$\|R^{(0)}(t,s)\|_{L^2 \to L^2} \leq C(t-s)^2$$

for $0 < |t-s| \leq \delta$.

We then conclude by applying Theorem 7.2.1 with $N = 1$.   □

8.5 Higher-Order Parametrices

In this section we increase the regularity level of the potential and accordingly prove a stronger convergence theorem. Namely, let $N \in \mathbb{N}$, $N \geq 1$. Suppose that

$V(t,x)$ satisfies Assumption (B') in Sect. 8.2 with $\kappa = d + 1 + N([d/2]+3)$.

Therefore we will apply the results of Sects. 8.2 and 8.3 for this value of κ.

Let δ be the constant appearing in Proposition 8.2.5, so that the function $S(t,s,x,y)$ is well defined for $0 < |t-s| \leq \delta$.

In order to manufacture higher-order parametrices, consider the functions $a_j(t,s,x,y)$, $j = 1, 2, \ldots, N$, defined by the formulas (cf. [108, Formulas (3.4) and (3.5)])

$$a_1(t,s,x,y) = \exp\left(-\frac{1}{2}\int_s^t (\tau - s)\Delta_x \omega(\tau,s,x(\tau),y)\,d\tau\right) \tag{8.32}$$

and

$$a_j(t,s,x,y) = -\frac{1}{2}a_1(t,s,x,y)\int_s^t \frac{\Delta_x a_{j-1}(\tau,s,x(\tau),y)}{a_1(\tau,s,x(\tau),y)}\,d\tau,$$

for $j = 2, \ldots, N$, where $\omega(t,s,x,y)$ is the function defined in (8.24) and $x(\tau) = x(\tau,s,y,\eta(t,s,x,y))$.

For future comparison we point out that they satisfy the transport equations

$$\frac{\partial a_k}{\partial t} + \sum_{j=1}^d \frac{\partial S}{\partial x_j}\frac{\partial a_k}{\partial x_j} + \frac{1}{2}\left(\Delta_x S - \frac{d}{t-s}\right)a_k = \frac{1}{2}\Delta_x a_{k-1},$$

with $a_0(t,s,x,y) \equiv 0$ and the initial conditions $a_1(s,s,x,y) = 1$ and $a_k(s,s,x,y) = 0$ for $k = 2, 3, \ldots$

We need the following result on the regularity of the functions $a_j(t,s,x,y)$.

Proposition 8.5.1 *For $0 < |t-s| \leq \delta$ we have*

$$a_1(t,s,x,y) = 1 - (t-s)^2 r(t,s,x,y) \tag{8.33}$$

and

$$1/a_1(t, s, x, y) = 1 + (t - s)^2 r'(t, s, x, y), \tag{8.34}$$

where the functions $r(t, s, x, y)$ and $r'(t, s, x, y)$ belong to a bounded subset of $H_{\mathrm{ul}}^{d+3+(N-1)([d/2]+3)}(\mathbb{R}^{2d})$.

Moreover, for $2 \le j \le N$ we have

$$\|a_j(t, s, \cdot, \cdot)\|_{H_{\mathrm{ul}}^{d+3+(N-j)([d/2]+3)}(\mathbb{R}^{2d})} \le C|t - s|^{j+1} \tag{8.35}$$

for some constant $C > 0$.

Proof Let us focus on (8.33). It follows from Proposition 8.3.1 (with $\kappa = d + 1 + N([d/2] + 3)$) that $\Delta_x \omega(\tau, s, x, y)$ belongs to a bounded subset of $H_{\mathrm{ul}}^{d+1+N([d/2]+3)}(\mathbb{R}^{2d})$ for $0 < |\tau - s| \le \delta$. Combining this fact with Proposition 3.6.6 (with the g replaced by the map $(y, \zeta) \mapsto (\tilde{x}(\tau, s, y, \zeta), y))$ we infer that the same conclusion holds for

$$\Delta_x \omega(\tau, s, \tilde{x}(\tau, s, y, \zeta), y),$$

where the function $\tilde{x}(\tau, s, y, \zeta)$ has been defined in Proposition 8.2.5.

By virtue of Proposition 3.6.4 we have that the functions

$$\Delta_x \omega(\tau, s, \tilde{x}(\tau, s, y, [(\tau - s)/(t - s)]\zeta), y)$$

belong, for $0 < |\tau - s| \le |t - s| \le \delta$, to a bounded subset of

$$H_{\mathrm{ul}}^{d+1+N([d/2]+3)-([d/2]+1)}(\mathbb{R}^{2d}) = H_{\mathrm{ul}}^{d+3+(N-1)([d/2]+3)}(\mathbb{R}^{2d}),$$

and again by Proposition 3.6.6 (with the map $(x, y) \mapsto (y, \zeta(t, s, x, y))$ in place of g) the same holds for

$$\Delta_x \omega(\tau, s, \underbrace{\tilde{x}(\tau, s, y, [(\tau - s)/(t - s)]\zeta(t, s, x, y)}_{=x(\tau)}, y) = \Delta_x \omega(\tau, s, x(\tau), y).$$

Hence we obtain the estimate

$$\left\| \int_s^t (\tau - s) \Delta_x \omega(\tau, s, x(\tau), y) \, d\tau \right\|_{H_{\mathrm{ul}}^{d+3+(N-1)([d/2]+3)}(\mathbb{R}^{2d})} \le C|t - s|^2.$$

After expanding the exponential in (8.32) as a power series in the Banach algebra $H_{\mathrm{ul}}^{d+3+(N-1)([d/2]+3)}(\mathbb{R}^{2d})$ we obtain (8.33), while (8.34) follows by the same arguments applied to $1/a_1(t, s, x, y)$.

To conclude, repeating the same path and arguing by induction on j yield that the functions $a_j(t, s, x, y)$ for $2 \leq j \leq N$ satisfy the estimates in (8.35). More precisely, for the case $j = 2$ one uses that $\Delta_x a_1(\tau, s, x, y)$ belongs to a bounded subset of $H_{ul}^{d+1+(N-1)([d/2]+3)}(\mathbb{R}^{2d})$, therefore $\Delta_x a_1(\tau, s, x(\tau), y)$ belongs to a bounded subset of $H_{ul}^{d+1+(N-1)([d/2]+3)-([d/2]+1)}(\mathbb{R}^{2d}) = H_{ul}^{d+3+(N-2)([d/2]+3)}(\mathbb{R}^{2d})$. The desired claim then follows by repeating this argument. $\square$

Let us now introduce the amplitude

$$a^{(N)}(t, s, x, y) = \sum_{j=1}^{N} \left(\frac{i}{\hbar}\right)^{1-j} a_j(t, s, x, y), \quad N = 1, 2, \ldots$$

and the corresponding oscillatory integral operator

$$E^{(N)}(t, s)f(x) = \frac{1}{(2\pi i(t-s)\hbar)^{d/2}} \int_{\mathbb{R}^d} e^{\frac{i}{\hbar} S(t,s,x,y)} a^{(N)}(t, s, x, y) f(y) \, dy.$$

It follows from Propositions 8.3.1, 8.5.1 and 4.4.9 that $E^{(N)}(t, s)$ is bounded in $L^2(\mathbb{R}^d)$ and satisfies the estimate

$$\|E^{(N)}(t, s)\|_{L^2 \to L^2} \leq C$$

for some constant $C > 0$, for $0 < |t - s| \leq \delta$, possibly for a smaller value of δ. Moreover, by Propositions 8.5.1 and 4.4.9 we have

$$\|E^{(N)}(t, s) - E^{(0)}(t, s)\|_{L^2 \to L^2} \leq C(t - s)^2,$$

so that, by Proposition 8.4.1, for every $f \in L^2(\mathbb{R}^d)$ we have

$$\lim_{t \to s} E^{(N)}(t, s)f = f$$

in $L^2(\mathbb{R}^d)$.

On the other hand we also have (cf. [108, Formulas (3.9) and (3.11)])

$$\left(i\hbar\partial_t + \frac{1}{2}\hbar^2\Delta - V(t, x)\right)E^{(N)}(t, s)f = G^{(N)}(t, s)f$$

with

$$G^{(N)}(t, s)f = -\frac{(-i\hbar)^{N+1}/2}{(2\pi i(t-s)\hbar)^{d/2}} \int_{\mathbb{R}^d} e^{\frac{i}{\hbar} S(t,s,x,y)} \Delta_x a_N(t, s, x, y)(t, s, x, y) f(y) \, dy.$$

By (8.33) (if $N = 1$) or (8.35) (if $N \geq 2$) we see that the amplitude $\Delta_x a_N$ satisfies the estimate

$$\|\Delta_x a_N(t, s, \cdot, \cdot)\|_{H_{ul}^{d+1}(\mathbb{R}^{2d})} \leq C|t - s|^{N+1}.$$

Proposition 4.4.9 thus gives

$$\|G^{(N)}(t, s)\|_{L^2 \to L^2} \leq C\hbar^{N+1}|t - s|^{N+1}. \tag{8.36}$$

Consider now the remainder operator

$$R^{(N)}(t, s)f := E^{(N)}(t, s)f - U(t, s)f = -\frac{i}{\hbar} \int_s^t U(t, \tau)G^{(N)}(\tau, s)f \, d\tau.$$

Using (8.36) we deduce that

$$\|R^{(N)}(t, s)\|_{L^2 \to L^2} \leq C\hbar^N|t - s|^{N+2}$$

for $0 < |t - s| \leq \delta$.

To conclude, consider the composition

$$E^{(N)}(\Omega, t, s) = E^{(N)}(t, t_{L-1})E^{(N)}(t_{L-1}, t_{L-2}) \ldots E^{(N)}(t_1, s),$$

for any subdivision $\Omega : s = t_0 < t_1 < \ldots < t_L = t$ of the interval $[s, t]$ such that $\omega(\Omega) \leq \delta$.

We have the following result.

Theorem 8.5.2 *Assume that $V(t, x)$ satisfies Assumption (B') in Sect. 8.2 with $\kappa = d + 1 + N([d/2] + 3)$, for some $N \geq 1$.*

For every $T > 0$ there exists $C = C(T) > 0$ such that for $0 < t - s \leq T$ and any subdivision Ω of the interval $[s, t]$ with $\omega(\Omega) \leq \delta$, and $0 < \hbar \leq 1$, we have

$$\|E^{(N)}(\Omega, t, s) - U(t, s)\|_{L^2 \to L^2} \leq C\hbar^N \omega(\Omega)^{N+1}(t - s).$$

Proof As in the case considered in Sect. 8.4 (which essentially corresponds to $N = 0$), the desired result follows from Theorem 7.2.1 (with N replaced by $N + 1$). □

Chapter 9
Convergence in the L^p Setting

9.1 Summary

The point of departure of this chapter is again the time-slicing approximation for Feynman path integrals discussed in Sect. 1.3.1 for the equation

$$i\hbar \partial_t \psi = -\frac{1}{2}\hbar^2 \Delta \psi + V(t,x)\psi.$$

We already stressed several times that the contributions by Fujiwara—in particular Theorem 1.3.1—provide a wide and deep analysis of the convergence problem in the standard L^2-based setting. In Chap. 8 we further explored the topic by embarking on a journey through several regularity assumptions on the potential, but still aiming at convergence in $\mathcal{L}(L^2(\mathbb{R}^d))$ for the time-slicing approximate propagators. We are now interested to derive convergence results in the L^p setting, with $p \neq 2$.

As already mentioned in Sect. 1.3.4, moving to this setting comes with an unavoidable loss of derivatives, since

$$\left\| e^{i\hbar \Delta/2} f \right\|_{L^p} \le C \left\| (1 - h\Delta)^{r/2} f \right\|_{L^p}, \quad r = 2d|1/2 - 1/p|, \quad 1 < p < \infty.$$

This estimate was proved in [177] in the case where $\hbar = 1$, whereas the general case follows at once by means of a scaling argument. Generalizations to different classes of potentials were proved, except for the endpoint, in several papers by exploiting in an essential way some smoothing effect (Gaussian estimates for the heat kernel), see for instance [20, 32, 38, 70, 152, 153]. The above loss of derivatives is optimal [29] and can be explained by the fact that the characteristic manifold of the Schrödinger operator, that is a paraboloid, has non-vanishing Gaussian curvature. For hyperbolic equations [69, 101, 196, 203, 212] we have instead the loss $r = (d-1)|1/2 - 1/p|$, because there is one flat direction in that case.

© The Author(s), under exclusive license to Springer Nature Switzerland AG 2022
F. Nicola, S. I. Trapasso, *Wave Packet Analysis of Feynman Path Integrals*,
Lecture Notes in Mathematics 2305, https://doi.org/10.1007/978-3-031-06186-8_9

On the basis of these observations, one is lead to resort to the scale of semiclassical L^p-based Sobolev spaces introduced in Chap. 5: recall that for $1 < p < \infty$ and
$r \in \mathbb{R}$ we defined

$$H_\hbar^{r,p}(\mathbb{R}^d) = \{f \in \mathcal{S}'(\mathbb{R}^d) \, : \, \|f\|_{H_\hbar^{r,p}} = \left\|(1 - h\Delta)^{r/2}f\right\|_{L^p} < \infty\}.$$

As demonstrated in [70], this framework is very well suited to the analysis of
Fourier integral operators originating from Schrödinger propagators associated with
quadratic Hamiltonian operators.

One is then confronted with a crucial issue: the space of bounded operators
$H_\hbar^{r,p} \to L^p$ is clearly not an algebra under composition. This is a major obstacle
for a proficient time-slicing approximation scheme, having in mind the construction
of the parametrices $E^{(N)}(\Omega, t, s)$ in Fujiwara's approach (cf. (1.19)).

The toolkit of Gabor analysis provides an efficient way to overcome this
difficulty, which becomes manageable as soon as one transfers the problem to the
phase space setting. Inspired by previous work in this spirit [81, 162, 172, 214]
and using the machinery developed in Chap. 4, we are able to exploit the sparsity
of the phase-space representations of the involved operators with respect to Gabor
wave packets. This approach leads to a convergence result in L^p, with peculiar
loss of derivatives, for Fujiwara's time-slicing approximate propagators in the same
regularity setting (cf. Assumption (A) in Sect. 1.3.1). We refer to Theorem 9.4.1
below for a detailed statement.

In short, the strategy of the proof goes as follows: one lifts the analysis to
the phase space level by means of non-trivial embeddings relating modulation
and Sobolev spaces [160, 161] (Theorem 5.1.1), then proves that the approximate
propagators in the form of oscillatory integral operators belong to suitable FIO
classes of operators (Definition 5.2.1)—hence well-behaved on modulation spaces
(Theorem 5.2.2). One should also keep track of $\hbar$ by means of suitable dilations, in
accordance with the semiclassical framework introduced in Chap. 5.

We emphasize the straightforwardness of this approach in comparison to the
painful analysis based on endpoint continuity results involving the Hardy space—
see for instance [81, 177, 203, 212]; all these technicalities are now hidden in the
above mentioned embedding and the analysis is focused at the level of phase space,
resulting in a remarkably simple and close to intuition handling of the matter.
Moreover, while here we confine ourselves to the smooth scenario as in Fujiwara
[108, 109], there is good reason to believe that the same approach may extend to
suitable classes of rough potentials. Some evidence of this issue is provided below.

We briefly describe the organization of this chapter. In Sect. 9.2 we review the
essential facts of the time-slicing approximation that will be used throughout the
discussion. In Sect. 9.3 we perform the transfer of the problem to the phase-space
level and prove that the relevant operators belong to suitable FIO classes, while
Sect. 9.4 is devoted to the proof of the main convergence result. A convergence result
in the same spirit but also in the presence of a suitable magnetic field (cf. [241]) is

stated and proved in Sect. 9.5. Finally, in Sect. 9.6 we discuss the sharpness of the results and in Sect. 9.7 we provide a partial extension to a class of rough potentials.

9.2   Review of the Short Time Analysis in the Smooth Category

Recall that the following condition is assumed to be satisfied by the potentials under our consideration.

Assumption (A) $V(t, x)$ *is a real-valued function of* $(t, x) \in \mathbb{R} \times \mathbb{R}^d$ *with* $\partial_x^\alpha V(t, x)$ *continuous in* $(t, x) \in \mathbb{R} \times \mathbb{R}^d$, *for every* $\alpha \in \mathbb{N}^d$ *and satisfying*

$$|\partial_x^\alpha V(t, x)| \leq C_\alpha, \quad |\alpha| \geq 2, \ (t, x) \in \mathbb{R} \times \mathbb{R}^d.$$

In particular the regularity estimates in Sects. 8.2 and 8.3 hold for every Sobolev exponent κ, and by Sobolev embedding we get corresponding estimates in the smooth category. Similarly, the analysis in Sect. 8.5 can be repeated at any Sobolev regularity (whereas there we set $\kappa = d + 1 + N([d/2] + 3)$). Let us review the estimates that one obtains—indeed, they were proved in [108, 109]; see also [241].

We consider the space $C_b^\infty(\mathbb{R}^{2d})$ of smooth function $a(x, y)$ in $\mathbb{R}^{2d}$ which are bounded together with their derivatives, endowed with the seminorms

$$\|a\|_m = \sup_{|\alpha|+|\beta| \leq m} \|\partial_x^\alpha \partial_y^\beta a\|_{L^\infty(\mathbb{R}^{2d})}, \quad m \in \mathbb{N}.$$

Consider now the Hamiltonian

$$H(t, x, \xi) = \frac{1}{2}|\xi|^2 + V(t, x).$$

Denote by $(x(t, s, y, \eta), \xi(t, s, y, \eta))$ $(s, t \in \mathbb{R}, \ y, \eta \in \mathbb{R}^d)$, the solution of the corresponding system

$$\dot{x} = \xi, \quad \dot{\xi} = -\nabla_x V(t, x)$$

with initial condition at time $t = s$ given by $x(s, s, y, \eta) = y, \xi(s, s, y, \eta) = \eta$. The flow

$$(x(t, s, y, \eta), \xi(t, s, y, \eta)) = \chi(t, s)(y, \eta)$$

defines a smooth canonical transformation $\chi(t,s)\colon \mathbb{R}^{2d} \to \mathbb{R}^{2d}$ satisfying for every $T_0 > 0$ the estimates

$$|\partial_y^\alpha \partial_\eta^\beta x(t,s,y,\eta)| + |\partial_y^\alpha \partial_\eta^\beta \xi(t,s,y,\eta)| \leq C_{\alpha,\beta}(T_0), \quad y,\eta \in \mathbb{R}^d, \tag{9.1}$$

for some constant $C_{\alpha,\beta}(T_0) > 0$, if $|t - s| \leq T_0$, $|\alpha| + |\beta| \geq 1$ ([109, Proposition 1.1]).

Moreover, there exists $\delta > 0$ such that for $0 < |t - s| \leq \delta$ and every $x, y \in \mathbb{R}^d$, there exists only one classical path γ such that $\gamma(s) = y$, $\gamma(t) = x$. By computing the action along this path γ, as in (1.15), we define the generating function $S(t,s,x,y)$ for $0 < |t - s| \leq \delta$. It satisfies the estimates

$$|t - s|\,|\partial_x^\alpha \partial_y^\beta S(t,s,x,y)| \leq C_{\alpha,\beta}, \quad |\alpha| + |\beta| \geq 2, \tag{9.2}$$

and

$$|t - s|\left|\det\left(\frac{\partial^2 S(t,s,x,y)}{\partial y^2}\right)\right| \geq \tilde{\delta}, \tag{9.3}$$

for some $\tilde{\delta} > 0$ and every $x, y \in \mathbb{R}^d$, always for $0 < |t - s| \leq \delta$.

We now recall the construction of the parametrices $E^{(N)}(t,s)$ from Sect. 8.5. For $0 < |t - s| \leq \delta$, define the operator $E^{(0)}(t,s)$ as in (8.2), namely

$$E^{(0)}(t,s)f(x) = \frac{1}{(2\pi i(t-s)\hbar)^{d/2}} \int_{\mathbb{R}^d} e^{\frac{i}{\hbar}S(t,s,x,y)} f(y)\,dy,$$

and for $N = 1, 2, \ldots$ set

$$E^{(N)}(t,s)f(x) = \frac{1}{(2\pi i(t-s)\hbar)^{d/2}} \int_{\mathbb{R}^d} e^{\frac{i}{\hbar}S(t,s,x,y)} a^{(N)}(\hbar,t,s,x,y) f(y)\,dy,$$

with $a^{(N)}(\hbar,t,s,x,y) = \sum_{k=1}^N (i\hbar)^{k-1} a_k(t,s,x,y)$, where the amplitudes a_k are defined in Sect. 8.5. We have, for every $m \in \mathbb{N}$,

$$\|a_k(t,s,\cdot,\cdot)\|_m \leq C_m \quad \text{for} \quad 0 < |t - s| \leq \delta. \tag{9.4}$$

The operators $E^{(N)}(t,s)$ are parametrices in the sense that, for $N = 0, 1, \ldots,$

$$\left(i\hbar\partial_t + \frac{1}{2}\hbar^2\Delta - V(t,x)\right) E^{(N)}(t,s)f = G^{(N)}(t,s)f$$

with

$$G^{(N)}(t,s)f = \frac{1}{(2\pi i(t-s)\hbar)^{d/2}} \int_{\mathbb{R}^d} e^{\frac{i}{\hbar}S(t,s,x,y)} g_N(\hbar,t,s,x,y) f(y)\,dy, \tag{9.5}$$

where g_N satisfies the estimates [109, Propositions 1.5, 1.6]

$$\|g_N(\hbar, t, s, \cdot, \cdot)\|_m \leq C_m \hbar^{N+1} |t - s|^{N+1}. \tag{9.6}$$

Moreover, for a subdivision $\Omega : s = t_0 < t_1 < \ldots < t_L = t$ with $\omega(\Omega) :=$ $\max\{t_j - t_{j-1}, \; j = 1, \ldots, L\} \leq \delta$ we define

$$E^{(N)}(\Omega, t, s) = E^{(N)}(t, t_{L-1}) E^{(N)}(t_{L-1}, t_{L-2}) \ldots E^{(N)}(t_1, s). \tag{9.7}$$

As already observed, if δ is small enough, for $0 < |t - s| \leq \delta$ the propagator has the form

$$U(t, s) f(x) = \frac{1}{(2\pi i (t - s)\hbar)^{d/2}} \int_{\mathbb{R}^d} e^{\frac{i}{\hbar} S(t,s,x,y)} b(\hbar, t, s, x, y) f(y) \, dy, \tag{9.8}$$

for an amplitude b such that $\partial_x^\alpha \partial_y^\beta b(\hbar, t, s, x, y)$ is of class C^1 in t, s and

$$\|b(\hbar, t, s, \cdot, \cdot)\|_m \leq C_m \tag{9.9}$$

for $0 < |t - s| \leq \delta, 0 < \hbar \leq 1, m \in \mathbb{N}$.

In the case where $|t - s|$ is large $U(t, s)$ coincides with the composition of a finite number of such oscillatory integral operators.

9.3   Wave Packet Analysis of the Schrödinger Flow

The main purpose of this section is to show that the oscillatory integral operators introduced above belong to the class $FIO_\hbar(\chi^h(t, s))$ defined in Definition 5.2.1, for a canonical transformation $\chi^h(t, s)$ defined as in (5.7). This fact will enable recourse to Theorem 5.2.4 whenever concerned with their composition for large time intervals.

We adhere to the notation of Sect. 9.2. In particular, the oscillatory integral operators are well defined for $0 < |t - s| \leq \delta$.

Proposition 9.3.1 *Let* $0 < |t - s| \leq \delta$ *and* $a \in C_b^\infty(\mathbb{R}^{2d})$. *Consider the operator*

$$Tf(x) = \frac{1}{(2\pi i (t - s)\hbar)^{d/2}} \int_{\mathbb{R}^d} e^{\frac{i}{\hbar} S(t,s,x,y)} a(x, y) f(y) \, dy.$$

Then T *can be written as a Fourier integral operator:*

$$Tf(x) = \int_{\mathbb{R}^d} e^{2\pi i h^{-1} \Phi(t,s,x,h\eta)} b(h, t, s, x, h\eta) \widehat{f}(\eta) \, d\eta, \tag{9.10}$$

where

$$\Phi(t, s, x, \eta) = y \cdot \eta + S(t, s, x, y)$$

is the generating function in the coordinates[1] x, η, and $b(h, t, s, \cdot, \cdot) \in C_b^\infty$. More precisely, for every $m \in \mathbb{N}$ there exists $m' \in \mathbb{N}$ such that

$$\|b(h, t, s, \cdot, \cdot)\|_m \le C \|a\|_{m'}$$

for some constant $C > 0$, depending only on m, the dimension d, upper bounds for a certain number of the derivatives of $(t - s)S(t, s, x, y)$ in (9.2) and the lower bound constant $\tilde{\delta}$ in (9.3).

Proof Set $\tilde{S}(t, s, x, y) = (t - s)S(t, s, x, y)$ and $\tilde{\eta} = h(t - s)\eta$. By the Plancherel theorem we can write T in the form

$$Tf(x) = \frac{1}{(2\pi i(t - s)\hbar)^{d/2}}$$

$$\times \int_{\mathbb{R}^d} \left(\int_{\mathbb{R}^d} \exp\left(i(\hbar(t - s))^{-1}[y \cdot \tilde{\eta} + \tilde{S}(t, s, x, y)] \right) a(x, y)\, dy \right) \widehat{f}(\eta)\, d\eta.$$

We now apply the stationary phase principle in the form proved in [8, page 320 and Lemma 3.2]; the assumptions are satisfied because for $0 < |t - s| \le \delta$ we have, by (9.2) and (9.3),

$$|\partial_x^\alpha \partial_y^\beta \tilde{S}(t, s, x, y)| \le C_{\alpha,\beta}, \quad |\alpha| + |\beta| \ge 2,$$

and

$$\left| \det\left(\frac{\partial^2 \tilde{S}(t, s, x, y)}{\partial y^2} \right) \right| \ge \tilde{\delta} > 0.$$

Moreover, since $y = y(t, s, x, \eta)$ is the unique solution of $-\eta = \partial S(t, s, x, y)/\partial y$, the function $y = y(t, s, x, \tilde{\eta}/(t - s))$ will be the unique solution to $-\tilde{\eta} = \partial \tilde{S}(t, s, x, y)/\partial y$, and we obtain

$$\frac{1}{(2\pi i(t - s)\hbar)^{d/2}} \int_{\mathbb{R}^d} \exp\left(i(\hbar(t - s))^{-1}[y \cdot \tilde{\eta} + \tilde{S}(t, s, x, y)] \right) a(x, y)\, dy$$

$$= \exp\left(i(\hbar(t - s))^{-1}[y(t, s, x, \tilde{\eta}/(t - s)) \cdot \tilde{\eta} + \tilde{S}(t, s, x, y(t, s, x, \tilde{\eta}/(t - s)))] \right)$$

$$\times b(\hbar, t, s, x, \tilde{\eta}))$$

---

[1] Namely, here $y = y(t, s, x, \eta)$ is the unique solution to $-\eta = \partial S(t, s, x, y)/\partial y$.

for some amplitude $b(\hbar, t, s, \cdot, \cdot)$ belonging to some bounded subset of C_b^∞, when $0 < |t - s| \le \delta, 0 < \hbar \le 1$. Again, we can write the last expression as

$$\exp\left(\frac{i}{\hbar}[y(t, s, x, h\eta) \cdot h\eta + S(t, s, x, y(t, s, x, h\eta))]\right) b(\hbar, t, s, x, (t - s)h\eta),$$

which gives the desired expression for Tf.

The estimates of the seminorms of b in terms of those of a follow from the proof of the stationary phase principle. $\square$

Corollary 9.3.2 *With the notation of Proposition 9.3.1, it turns out that $T \in FIO_\hbar(\chi^h(t, s))$, with*

$$\chi^h(t, s)(y, \eta) = h^{-1/2}\chi(t, s)(h^{1/2}y, h^{1/2}\eta).$$

Moreover, for every $m \in \mathbb{N}$ there exists $m' \in \mathbb{N}$ such that, for $0 < |t - s| \le \delta$,

$$\|T\|_{m, \chi^h(t,s)}^\hbar \le C\|a\|_{m'}$$

for some constant $C > 0$ independent of a and $\hbar$, t, s $(0 < \hbar \le 1, 0 < |t - s| \le \delta)$.

Proof We have to prove that $D_{h^{1/2}}T D_{h^{-1/2}} \in FIO(\chi^h(t, s))$. Using (9.10) and a scaling argument we can write

$$D_{h^{1/2}}T D_{h^{-1/2}} f(x) = \int_{\mathbb{R}^d} e^{2\pi i h^{-1}\Phi(t, s, h^{1/2}x, h^{1/2}\eta)} b(h, t, s, h^{1/2}x, h^{1/2}\eta) \widehat{f}(\eta) \, d\eta.$$

Now, the phase $h^{-1}\Phi(t, s, h^{1/2}x, h^{1/2}\eta)$ generates the canonical transformation $\chi^h(t, s)$ as in the statement. Moreover, $h^{-1}\Phi(t, s, h^{1/2}x, h^{1/2}\eta)$ is tame *uniformly* with respect to h and t, s, for $0 < h \le 2\pi$ and $0 < |t - s| \le \delta$, in the sense that the required bounds hold with constants independent of these parameters, or equivalently $\chi^h(t, s)$ satisfies the properties *A1,A2,A3* in Sect. 4.4.1, with uniform bounds. This is clear, because $\chi(t, s)$ satisfies *A1,A2,A3* uniformly with respect to t, s for $0 < |t - s| \le \delta$ (possibly for a smaller value of δ): *A1,A2* follow from (9.1), whereas *A3* holds by virtue of [241, Proposition 2.3'] applied with $\alpha = \beta = 0$, even in the presence of a magnetic field as in Sect. 9.5 below.

Finally, the symbol $b(h, t, s, h^{1/2}x, h^{1/2}\eta)$ has seminorms in $C_b^\infty(\mathbb{R}^{2d})$ dominated by those of $b(h, t, s, x, \eta)$. The claim thus follows directly from Theorem 4.4.4. $\square$

Remark 9.3.3 With the notation of the previous result, notice that for any $s, \tau, t \in \mathbb{R}$ we have

$$\chi^h(t, s) = \chi^h(t, \tau) \circ \chi^h(\tau, s).$$

Indeed, this follows at once from the case $h = 1$ and the definition of χ^h.

9.4   Convergence in L^p with Loss of Derivatives

With the notation of Sect. 9.2, consider any subdivision $\Omega : s = t_0 < t_1 < \ldots < t_L = t$ with $\omega(\Omega) := \max\{t_j - t_{j-1}, \ j = 1, \ldots, L\} \leq \delta$ and set

$$E^{(N)}(\Omega, t, s) = E^{(N)}(t, t_{L-1}) E^{(N)}(t_{L-1}, t_{L-2}) \ldots E^{(N)}(t_1, s).$$

Theorem 9.4.1 *Assume the condition in Assumption (A). Let* $1 < p < \infty$ *and* $r = 2d|1/2 - 1/p|$.

(i) *For every* $T_0 > 0$ *there exists a constant* $C(T_0) > 0$ *such that, for all* $f \in \mathcal{S}(\mathbb{R}^d)$, $|s - t| \leq T_0$, $0 < \hbar \leq 1$,

$$\|U(t, s) f\|_{L^p} \leq C(T_0) \|f\|_{H_\hbar^{r,p}}, \quad 1 < p \leq 2, \tag{9.11}$$

$$\|U(t, s) f\|_{H_\hbar^{-r,p}} \leq C(T_0) \|f\|_{L^p}, \quad 2 \leq p < \infty. \tag{9.12}$$

(ii) *For every* $T_0 > 0$, $N = 0, 1, 2, \ldots$, *there exists a constant* $C(T_0) > 0$ *such that, for* $0 < t - s \leq T_0$ *and any sufficiently fine subdivision* Ω *of the interval* $[s, t]$, $f \in \mathcal{S}(\mathbb{R}^d)$, $0 < \hbar \leq 1$, *we have*

$$\left\|\left(E^{(N)}(\Omega, t, s) - U(t, s)\right) f\right\|_{L^p} \leq C(T_0) \hbar^N \omega(\Omega)^{N+1} (t - s) \|f\|_{H_\hbar^{r,p}},$$
$$\tag{9.13}$$

for $1 < p \leq 2$, *and*

$$\left\|\left(E^{(N)}(\Omega, t, s) - U(t, s)\right) f\right\|_{H_\hbar^{-r,p}} \leq C(T_0) \hbar^N \omega(\Omega)^{N+1} (t - s) \|f\|_{L^p},$$
$$\tag{9.14}$$

for $2 \leq p < \infty$.

Proof Let us first prove (9.11) and (9.12). We know from Sect. 9.2 that the propagator $U(t, s)$ is an oscillatory integral operator of the form (9.8) for $0 < |t - s| \leq \delta$, and therefore by Corollary 9.3.2 we have $U(t, s) \in FIO_\hbar(\chi^h(t, s))$, with seminorms $\|U(t, s)\|_{m, \chi^h(t,s)}^\hbar$, $m \in \mathbb{N}$, uniformly bounded with respect to $\hbar, t, s$ for $0 < |t - s| \leq \delta$. Using the evolution properties of the propagator, Remark 9.3.3 and Theorem 5.2.4 we see that, for any $T_0 > 0$ we have $U(t, s) \in FIO_\hbar(\chi^h(t, s))$ with seminorms satisfying

$$\|U(t, s)\|_{m, \chi^h(t,s)}^\hbar \leq C_0 \tag{9.15}$$

for a constant C_0 independent of $\hbar, t, s$, for $|t - s| \le T_0$ (but depending on T_0); here we are using the fact that the constant C in Theorem 5.2.4 can be chosen independent of such parameters, being $\chi^h(t, s)$ uniformly tame. We then deduce (9.11) and (9.12) from Corollary 5.2.3.

We now prove (9.13) and (9.14). By arguing as above it suffices to prove that

$$\| E^{(N)}(\Omega, t, s) - U(t, s) \|_{m, \chi^h(t,s)}^{\hbar} \le C(T_0) \hbar^N \omega(\Omega)^{N+1} (t - s) \tag{9.16}$$

for $0 < t - s \le T_0$ and for some $m > 2d$.

Now, by Corollary 9.3.2 and (9.6) the operator $G^{(N)}(t, s)$ defined in (9.5) belongs to $FIO_\hbar(\chi^h(t, s))$, with seminorms satisfying

$$\| G^{(N)}(t, s) \|_{m, \chi^h(t,s)}^{\hbar} \le C_m \hbar^{N+1} (t - s)^{N+1} \tag{9.17}$$

for a constant C_m independent of $\hbar, t, s$, for $0 < |t - s| \le \delta$. On the other hand, we have

$$R^{(N)}(t, s) f := E^{(N)}(t, s) f - U(t, s) f = -\frac{i}{\hbar} \int_s^t U(t, \tau) G^{(N)}(\tau, s) f \, d\tau,$$

so that by (9.15), (9.17) and Theorem 5.2.4 we obtain

$$\| R^{(N)}(t, s) \|_{m, \chi^h(t,s)}^{\hbar} \le C_m \hbar^N (t - s)^{N+2}. \tag{9.18}$$

Then, we can write

$$E^{(N)}(\Omega, t, s) - U(t, s)$$
$$= \left(U(t, t_{L-1}) + R^{(N)}(t, t_{L-1}) \right) \ldots \left(U(t_1, s) + R^{(N)}(t_1, s) \right) - U(t, s)$$

and argue as in Theorem 7.2.1 (cf. [109, Lemma 3.2]), now with the seminorms $\| \cdot \|_{m, \chi^h(t,s)}^{\hbar}$ (therefore depending on the time interval) in place of the L^2 norm. The composition estimate in Theorem 5.2.4 will play a key role in this program. We sketch the argument for the benefit of the reader.

Expanding the product above gives a sum of ordered products of operators, each term of such product having the following structure: *from right to left* we have, say, q_1 factors of type U, p_1 factors of type $R^{(N)}$, q_2 factors of type U, p_2 factors of type $R^{(N)}$, etc., up to q_k factors of type U, p_k factors of type $R^{(N)}$, to finish with q_{k+1} factors of type U. Here $p_1, \ldots, p_k, q_1, \ldots q_k, q_{k+1}$ are non-negative integers whose sum is L, with $p_j > 0$ and we can of course group together the consecutive factors of type U, using the evolution property of the propagator. We estimate the seminorm $\| \cdot \|_{m, \chi^h(t,s)}^{\hbar}$ of such an ordered product, for some fixed $m > 2d$, using a combination o Theorem 5.2.4 and the already derived estimates for each factor, namely (9.15) and (9.18), which we conveniently recast here as

$$\| R^{(N)}(t_j, t_{j-1})\|_{m,\chi^h(t_j,t_{j-1})}^\hbar \leq \tilde{C}_m \hbar^N (t_j - t_{j-1})^{N+2}.$$

Let C be the constant in Theorem 5.2.4. We can thus bound the seminorm $\| \cdot \|_{m,\chi^h(t,s)}^\hbar$ of the previous ordered product by

$$C^{p_1+\dots+p_k+k} C_0^{k+1} \prod_{j=1}^{k} \prod_{i=1}^{p_j} \tilde{C}_m \hbar^N (t_{J_j+i} - t_{J_j+i-1})^{N+2},$$

where C_0 comes from (9.15), $J_j = p_1 + \dots + p_{j-1} + q_1 + \dots + q_j$ for $j \geq 2$ and $J_1 = q_1$. In the last part of the proof of 7.2.1 we already showed that the sum over $p_1, \dots, p_k, q_1, \dots, q_{k+1}$ of these terms is in turn dominated by

$$C(T_0) \hbar^N \omega(\Omega)^{N+1} (t - s)$$

for $0 < t - s \leq T_0$. This gives (9.16) and concludes the proof.   □

9.5 The Case of Magnetic Fields

The purpose of this section is to extend some of the previous results in the presence of a magnetic field. To this aim, consider the Schrödinger equation

$$i\hbar \partial_t \psi = \frac{1}{2}\big(-i\hbar\nabla - A(t,x)\big)^2 \psi + V(t,x)\psi$$

where $V(t,x)$ and $A(t,x) = (A_1(t,x), \dots, A_d(t,x))$ are electric scalar and magnetic vector potential of the field, $t \in \mathbb{R}$, $x \in \mathbb{R}^d$.

Assume the following hypothesis (cf. [110, 226, 241]).

Assumption (B)

(a) *For $j = 1, \dots, d$, $A_j(t,x)$ is a real function of $(t,x) \in \mathbb{R} \times \mathbb{R}^d$ and $\partial_x^\alpha A_j(t,x)$ is C^1 in $(t,x) \in \mathbb{R} \times \mathbb{R}^d$, for every $\alpha \in \mathbb{N}^d$. Moreover, there exists $\epsilon > 0$ such that*

$$|\partial_x^\alpha B(t,x)| \leq C_\alpha (1 + |x|)^{-1-\epsilon}, \qquad |\alpha| \geq 1,$$

$$|\partial_x^\alpha A(t,x)| + |\partial_x^\alpha \partial_t A(t,x)| \leq C_\alpha, \qquad |\alpha| \geq 1,$$

for $(t,x) \in \mathbb{R} \times \mathbb{R}^d$, where $B(t,x)$ is the magnetic field, i.e. the skew-symmetric matrix with entries $B_{j,k}(t,x) = (\partial A_k/\partial x_j - \partial A_j/\partial x_k)(t,x)$.

(b) $V(t, x)$ *is a real function of* $(t, x) \in \mathbb{R} \times \mathbb{R}^d$ *with* $\partial_x^\alpha V(t, x)$ *continuous in* $(t, x) \in \mathbb{R} \times \mathbb{R}^d$, *for every* $\alpha \in \mathbb{N}^d$ *and satisfying*

$$|\partial_x^\alpha V(t, x)| \le C_\alpha, \quad |\alpha| \ge 2, \ (t, x) \in \mathbb{R} \times \mathbb{R}^d.$$

When $V = 0$ and under Assumption (B) it was proved in [241, Sections 2,3] that all the results concerning the propagator that were valid for short times, as well as the construction of the operators $E^{(N)}(t, s)$, $G^{(N)}(t, s)$ summarized in Sect. 9.2, extend to the case where $N \ge 1$. To be more precise, the same formulas for $E^{(N)}(t, s)$, $G^{(N)}(t, s)$ and $E^{(N)}(\Omega, t, s)$ hold, where now the amplitudes a_k satisfy the transport equations

$$\frac{\partial a_k}{\partial t} + \sum_{j=1}^{d} \left(\frac{\partial S}{\partial x_j}(t, s, x, y) - A_j(t, x) \right) \frac{\partial a_k}{\partial x_j}$$
$$+ \frac{1}{2}\left(\Delta_x S(t, s, x, y) - \frac{d}{t - s} - \mathrm{div}_x\, A(t, x) \right) a_k = \frac{1}{2}\Delta_x a_{k-1}, \qquad (9.19)$$

with $a_0(t, s, x, y) \equiv 0$ and the initial conditions $a_1(s, s, x, y) = 1$ and $a_k(s, s, x, y) = 0$ for $k = 1, 2, \ldots$. These transport equations hold even when the electric potential is present. Indeed, since S satisfies the Hamilton-Jacobi equation $\partial_t S + \frac{1}{2}(\nabla_x S - A)^2 + V = 0$, we have

$$\left(i\hbar\partial_t - \frac{1}{2}(-i\hbar\nabla_x - A)^2 - V(t, x) \right) \frac{1}{(2\pi i\hbar(t-s))^{d/2}} \int_{\mathbb{R}^d} e^{\frac{i}{\hbar}S(t,s,x,y)} a(x, y)\, dy$$

$$= \frac{1}{(2\pi i\hbar(t - s))^{d/2}}$$
$$\times \int_{\mathbb{R}^d} e^{\frac{i}{\hbar}S(t,s,x,y)} i\hbar \left[\partial_t a + (\nabla_x S - A)\cdot\nabla_x a + \frac{1}{2}\left(\Delta_x S - \frac{d}{t - s} - \mathrm{div}_x\, A\right) - \frac{i\hbar}{2}\Delta_x a \right] dy.$$

The key formulas (9.4), (9.5), (9.6), (9.7), (9.8) and (9.9) are still valid, for $N \ge 1$. As a consequence, we obtain the following result.

Theorem 9.5.1 *Under the Assumption* (B) *above, the conclusions of Theorem 9.4.1 still hold, at least for* $N \ge 1$.

Proof We have already observed that the results in Sect. 9.2 still hold under the Assumption (B), if $V = 0$ and $N \ge 1$, therefore the whole subsequent analysis applies, giving the desired result if $V = 0$.

Let us examine the case when a potential $V(t, x)$ is present. We prove only part *(ii)* of Theorem 9.4.1, the first part being similar and actually easier. Consider the gauge transformation:

$$G(t)u(t, x) = \exp \frac{i}{\hbar} G(t, x)u(t, x),$$

mapping the solution of the equation with potentials A, V into the solution of the equation with gauge potentials $A + \nabla_x G, V - \partial_t G$, cf. [241]. To be precise, if $U(t, s)$ and $U'(t, s)$ are the corresponding propagators, with the prime denoting that of the new equation, we have

$$U(t, s) = G(t)^{-1} U'(t, s) G(s).$$

In particular, if we choose $G(t, x) = \int_0^t V(\tau, x) d\tau$, in the new equation the electric potential is absent and magnetic potential is $A' = A + \int_0^t \partial_x V(\tau, x) d\tau$, still satisfying the above Assumption (B). Since the Lagrangian function $\mathcal{L}(x, v, t) = \frac{1}{2}|v|^2 + A(t, x) \cdot v - V(t, x)$ changes to $\mathcal{L}' = \mathcal{L} + dG(t, x(t))/dt$, the flow remains the same and the new generating function is

$$S'(t, s, x, y) = S(t, s, x, y) + \int_0^t V(\tau, x) d\tau - \int_0^s V(\tau, y) d\tau,$$

where $S(t, s, x, y)$ is the generating function for the original equation. The new amplitudes coincide with those corresponding to the original equation, since the transport equations (9.19) are invariant under the substitution $A \to A + \int_0^t \nabla_x V(\tau, x) d\tau, S \to S', V \to 0$. We thus infer

$$E^{(N)}(t, s) = G^{-1}(t) E^{(N)'}(t, s) G(s)$$

and therefore

$$E^{(N)}(\Omega, t, s) = G^{-1}(t) E^{(N)'}(\Omega, t, s) G(s),$$

where the primes are used to denote the corresponding operators for the new equation. As a result,

$$E^{(N)}(\Omega, t, s) - U(t, s) = G^{-1}(t) \Big(E^{(N)'}(t, s) - U'(t, s) \Big) G(s).$$

Given that (9.16) holds true for the difference $E^{(N)'}(\Omega, t, s) - U'(t, s)$, we prove now the same estimate for $E^{(N)}(\Omega, t, s) - U(t, s)$. By virtue of Theorem 5.2.4, it suffices to prove that $G(t) \in FIO_\hbar(\tilde{\chi}^h)$ for some tame canonical transformation $\tilde{\chi}$. This follows by Corollary 9.3.2, because $G(t)$ can be written in the form (9.10) with $\Phi(t, x, \eta) = x\eta + \int_0^t V(\tau, x) d\tau, b = 1$, and

$$\tilde{\chi}(y, \eta) = \Big(y, \eta + \int_0^t \nabla_x V(\tau, x) d\tau \Big).$$

$\square$

Remark 9.5.2 The question whether the conclusion of Theorem 9.4.1 holds for $N = 0$ in presence of a magnetic field as above (in [241, Theorem 5] the case $N = 0$ is excluded as well) is open. Indeed, the approach in [109] requires very precise estimates for the generating function $S(t, s, x, y)$, whose derivation in presence of a magnetic field is a definitely non-trivial issue.

9.6 Sharpness of the Results

An aspect definitely worthy of mention is that the dichotomy $p > 2$, $p < 2$ in Theorem 9.4.1 is unavoidable. In support of this claim, consider the case of the harmonic oscillator

$$\frac{i}{2\pi}\partial_t \psi = -\frac{1}{8\pi^2}\Delta\psi + \frac{1}{2}|x|^2\psi, \tag{9.20}$$

where we take $\hbar = (2\pi)^{-1}$ for simplicity—so that $H_\hbar^{r,p} = H^{r,p}$. By Mehler's formula in Example 4.3.10 the propagator $U(t, 0)$ at $t = \pi/2$ coincides (up to a phase factor) with the Fourier transform $f \mapsto \hat{f}$.

Proposition 9.6.1 *Let* $1 < p < \infty$, $r_1, r_2 \in \mathbb{R}$. *The Fourier transform maps* $H^{r_1,p}(\mathbb{R}^d) \to H^{r_2,p}(\mathbb{R}^d)$ *continuously if and only if*

$$r_1 \geq 2d(1/p - 1/2) \text{ and } r_2 \leq 0, \quad 1 < p \leq 2$$

$$r_1 \geq 0 \text{ and } r_2 \leq -2d(1/2 - 1/p), \quad 2 \leq p < \infty.$$

Proof (Sufficient Conditions) The desired estimates can be inferred using the embeddings of Sobolev spaces and duality arguments once the boundedness of the Fourier transform $H^{r,p}(\mathbb{R}^d) \to L^p(\mathbb{R}^d)$ with $r = 2d(1/p - 1/2)$, for $1 < p \leq 2$, holds. This is actually equivalent to the estimate

$$\|(1 + 4\pi^2|\xi|^2)^{-r/2}\hat{f}(\xi)\|_{L^p} \leq C\|f\|_{L^p},$$

which is in turn a direct consequence of the Hardy-Littlewood-Paley inequality [18, Theorem 1.4.1]—that is the same estimate with $(1 + 4\pi^2|\xi|^2)^{-r/2}$ replaced by $|\xi|^{-r}$ in the left-hand side.

Of course, the result also follows from Theorem 9.4.1 applied to Eq. (9.20), with $s = 0, t = \pi/2$ ($\hbar = (2\pi)^{-1}$).

Necessary conditions. Let us first prove the condition on r_2 when $2 \leq p < \infty$. By duality, this will give the condition on r_1 for $1 < p \leq 2$ as well.

Consider the space

$$\mathcal{A} = \{f \in \mathcal{S}(\mathbb{R}^d) : \hat{f}(\xi) = 0 \text{ for } |\xi| \geq 1\}.$$

By Bernstein inequalities (see e.g. [239, Proposition 5.3]) there exists a constant $C > 0$ such that

$$\|f\|_{H^{r_1,p}} \leq C\|f\|_{L^p}, \quad \forall f \in \mathcal{A},$$

even for $r_1 < 0$ (for $r_1 \geq 0$ this is trivially true for every $f \in S(\mathbb{R}^d)$).

Hence, if the Fourier transform is bounded $H^{r_1,p} \to H^{r_2,p}$, we have

$$\|\widehat{f}\|_{H^{r_2,p}} \leq C\|f\|_{L^p}, \quad \forall f \in \mathcal{A}.$$

Now, suppose by contradiction that $r_2 > -2d(1/2 - 1/p)$. Fix $f \in \mathcal{A} \setminus \{0\}$ and test this estimate on $f(x/\lambda)$, with $\lambda \geq 1$, so that $f(x/\lambda)$ belongs to $\mathcal{A}$ too. We have

$$\|(1 - \Delta)^{r_2/2} \widehat{f(\cdot/\lambda)}\|_{L^p} \leq C\|f(\cdot/\lambda)\|_{L^p} \tag{9.21}$$

where

$$(1 - \Delta)^{r_2/2} \widehat{f(\cdot/\lambda)} = \lambda^{d+r_2}[(\lambda^{-2} - \Delta)^{r_2/2} \widehat{f}](\lambda \cdot).$$

By the dominated convergence theorem ($|x|^{r_2}$ is locally integrable, because $r_2 > -d$) we have $[(\lambda^{-2} - \Delta)^{r_2/2} \widehat{f}](x) \to |2\pi D|^{r_2} \widehat{f}(x)$ for every $x \in \mathbb{R}^d$, and therefore by the Fatou lemma,

$$0 \neq \||2\pi D|^{r_2} \widehat{f}\|_{L^p} \leq \liminf_{\lambda \to +\infty} \|(\lambda^{-2} - \Delta)^{r_2/2} \widehat{f}\|_{L^p}.$$

Hence, letting $\lambda \to +\infty$ in (9.21) we obtain $r_2 \leq -2d(1/2 - 1/p)$, which is a contradiction.

Let us now prove that $r_2 \leq 0$ if $1 < p \leq 2$. This will give the condition on r_1 for $2 \leq p < \infty$ as well.

Fix $f \in S(\mathbb{R}^d) \setminus \{0\}$, and let $f_\lambda(x) = f(x_1 - \lambda, x_2, \ldots, x_d)$, $\lambda > 0$. Suppose that the following estimate holds:

$$\|\widehat{f_\lambda}\|_{H^{r_2,p}} \leq C\|f_\lambda\|_{H^{r_1,p}} = C\|f\|_{H^{r_1,p}}. \tag{9.22}$$

We have

$$(1 - \Delta)^{r_2/2} \widehat{f_\lambda}(\xi)$$

$$= (1 - \Delta)^{r_2/2}[e^{-2\pi i \lambda \xi_1} \widehat{f}(\xi)]$$

$$= \int_{\mathbb{R}^d} e^{2\pi i x \cdot \xi}(1 + 4\pi^2 |x|^2)^{r_2/2} f(-(x_1 + \lambda), -x_2, \ldots, -x_d) \, dx$$

$$= \lambda^{r_2} e^{-2\pi i \lambda \xi_1} \int_{\mathbb{R}^d} e^{-2\pi i y \cdot \xi}$$

$$\times (\lambda^{-2} + 4\pi^2 |1 + \lambda^{-1} y_1|^2 + 4\pi^2 \lambda^{-2} |y_2|^2 + \ldots + 4\pi^2 \lambda^{-2} |y_d|^2)^{k_2/2} f(y) \, dy.$$

As $\lambda \to +\infty$ the last integral converges to $(2\pi)^{r_2} \widehat{f}(\xi)$, for every $\xi \in \mathbb{R}^d$, by the dominated convergence theorem. One can then conclude as above, after multiplying (9.22) by λ^{-r_2}, letting $\lambda \to +\infty$ and using the Fatou lemma.   □

9.7 Extensions to the Case of Rough Potentials

We finally discuss an extension of the first part of Theorem 9.4.1 to potentials in the Sjöstrand class

$$M^{\infty,1}(\mathbb{R}^d) = \left\{ f \in S'(\mathbb{R}^d) : \|f\|_{M^{\infty,1}(\mathbb{R}^d)} := \int_{\mathbb{R}^d} \|V_g f(\cdot, \xi)\|_{L^\infty} \, d\xi < \infty \right\},$$

see Sect. 3.5, where $g \in S(\mathbb{R}^d) \setminus \{0\}$, as usual.

We recall that functions in this space are bounded in $\mathbb{R}^d$ and locally have the mild regularity of a function whose Fourier transform is in L^1. In particular, there is no a priori condition on the existence of derivatives.

Consider now the equation (with $\hbar = 1$, for simplicity)

$$i \partial_t \psi = (-\Delta)^{k/2} \psi + V_2(t, x) \psi + V_1(t, x) \psi + V_0(t, x) \psi$$

with real $0 < k \le 2$, $t \in \mathbb{R}$, $x \in \mathbb{R}^d$. Suppose

$$\partial_x^\alpha V_j(t, \cdot) \in M^{\infty,1}(\mathbb{R}^d) \text{ for } |\alpha| = j, \ j = 0, 1, 2,$$

with V_2 and V_1 real-valued. Assume moreover that the map $\mathbb{R} \ni t \mapsto \partial_x^\alpha V_j(t, \cdot) \in M^{\infty,1}(\mathbb{R}^d)$ for $|\alpha| = j$, $j = 0, 1, 2$, are continuous in the sense of the narrow convergence; see Sect. 3.6.5.

Under this assumption, the propagator $U(t, s)$ was constructed in [58] and shown to be bounded $M^p(\mathbb{R}^d) \to M^p(\mathbb{R}^d)$ for every $1 \le p \le \infty$. As a consequence of Theorem 3.2.6 we then see that $U(t, s)$ enjoys the continuity property

$$U(t, s) : H^{r,p}(\mathbb{R}^d) \to L^p(\mathbb{R}^d), \quad 1 < p \le 2,$$

$$U(t, s) : L^p(\mathbb{R}^d) \to H^{-r,p}(\mathbb{R}^d), \quad 2 \le p < \infty,$$

with $r = 2d|1/2 - 1/p|$.

Bibliography

1. S. Albeverio, Z. Brzeźniak, Finite-dimensional approximation approach to oscillatory integrals and stationary phase in infinite dimensions. J. Funct. Anal. **113** (1), 177–244 (1993)
2. S. Albeverio, R. Høegh-Krohn, Oscillatory integrals and the method of stationary phase in infinitely many dimensions, with applications to the classical limit of quantum mechanics. I. Invent. Math. **40**(1), 59–106 (1977)
3. S. Albeverio, S. Mazzucchi, A unified approach to infinite- dimensional integration. Rev. Math. Phys. **28**(2), 1650005, 43 (2016)
4. S. Albeverio, P. Blanchard, R. Høegh-Krohn, Feynman path integrals and the trace formula for the Schrödinger operators. Comm. Math. Phys. **83** (1), 49–76 (1982)
5. S. Albeverio, R. Høegh-Krohn, S. Mazzucchi, *Mathematical Theory of Feynman Path Integrals*, 2nd edn., vol. 523 (Springer, Berlin, 2008)
6. J.-P. Antoine, Quantum mechanics beyond Hilbert space, in *Irreversibility and Causality (Goslar 1996)*, vol. 504 (Springer, Berlin, 1998), pp. 3–33
7. G. Arsu, On Kato-Sobolev type spaces (2013). arXiv: 1209.6465
8. K. Asada, D. Fujiwara, On some oscillatory integral transfor mations in $L^2(\mathbf{R}^n)$. Jpn. J. Math. **4**(2), 299–361 (1978)
9. G. Ascensi, Y. Lyubarskii, K. Seip, Phase space distribution of Gabor expansions. Appl. Comput. Harmon. Anal. **26**(2), 277–282 (2009)
10. V. Bargmann, P. Butera, L. Girardello, J.R. Klauder, On the completeness of the coherent states. Rep. Math. Phys. **2**(4), 221–228 (1971)
11. D. Bayer, Bilinear time-frequency distributions and pseudodifferential operators, Ph.D. thesis. University of Vienna (2010)
12. D. Bayer, E. Cordero, K. Gröchenig, S.I. Trapasso, Linear perturbations of the Wigner transform and the Weyl quantization, in *Advances in Microlocal and Time-Frequency Analysis* (Birkhäuser, Basel, 2020), pp. 79–120
13. A. Benedek, R. Panzone, The space L^p with mixed norm. Duke Math. J. **28**, 301–324 (1961)
14. C. Bennett, R. Sharpley, *Interpolation of Operators*, vol. 129 (Academic Press Inc., Boston, 1988)
15. Á. Bényi, K. Gröchenig, K.A. Okoudjou, L.G. Rogers, Unimodular Fourier multipliers for modulation spaces. J. Funct. Anal. **246**(2), 366–384 (2007)
16. Á. Bényi, K.A. Okoudjou, *Modulation Spaces* (Birkhäuser, New York, 2020)
17. F.A. Berezin, M.A. Shubin, Symbols of operators and quantization, in *Hilbert Space Operators and Operator Algebras (Proc. Internat. Conf., Tihany 1970)* (1972), pp. 21–52. Colloq. Math. Soc. János Bolyai, vol. 5
18. J. Bergh, J. Löfström, *Interpolation Spaces. An Introduction* (Springer, Berlin, 1976)

© The Author(s), under exclusive license to Springer Nature Switzerland AG 2022
F. Nicola, S. I. Trapasso, *Wave Packet Analysis of Feynman Path Integrals*,
Lecture Notes in Mathematics 2305, https://doi.org/10.1007/978-3-031-06186-8

19. G.D. Birkhoff, Quantum mechanics and asymptotic series. Bull. Amer. Math. Soc. **39**(10), 681–700 (1933)
20. S. Blunck, Generalized Gaussian estimates and Riesz means of Schrödinger groups. J. Aust. Math. Soc. **82**(2), 149–162 (2007)
21. B. Boashash (ed.), *Time-Frequency Signal Analysis and Processing* (Academic Press (Elsevier), Cambridge, 2015)
22. P. Boggiatto, E. Carypis, A. Oliaro, Wigner representations associated with linear transformations of the time-frequency plane, in *Pseudo-Differential Operators: Analysis, Applications and Computations*, vol. 213 (Birkhäuser/Springer Basel AG, Basel, 2011), pp. 275–288
23. P. Boggiatto, G. De Donno, A. Oliaro, Time-frequency representations of Wigner type and pseudo-differential operators. Trans. Amer. Math. Soc. **362**(9), 4955–4981 (2010)
24. P. Boggiatto, G. De Donno, A. Oliaro, B.K. Cuong, Generalized spectrograms and T-Wigner transforms. Cubo **12**(3), 171–185 (2010)
25. N. Bohr, *Niels Bohr – Collected Works*, vol. 3. The Correspondence Principle (1918–1923) (Elsevier/North-Holland, Amsterdam, 1976)
26. M. Born, P. Jordan, Zur Quantenmechanik. Z. Phys. **34**, 858–888 (1925)
27. A. Boulkhemair, Estimations L^2 précisées pour des intégrales oscillantes. Comm. Partial Differ. Equ. **22**(1–2), 165–184 (1997)
28. A. Boulkhemair, Remarks on a Wiener type pseudodifferential algebra and Fourier integral operators. Math. Res. Lett. **4**(1), 53–67 (1997)
29. P. Brenner, V. Thomée, L.B. Wahlbin, *Besov Spaces and Applications to Difference Methods for Initial Value Problems* (Springer, Berlin, 1975)
30. L.M. Brown (ed.), *Feynman's Thesis* (World Scientific Publishing Co. Pte. Ltd., Hackensack, 2005)
31. R.C. Busby, H.A. Smith, Product-convolution operators and mixed-norm spaces. Trans. Amer Math. Soc. **263**(2), 309–341 (1981)
32. F. Cacciafesta, P. D'Ancona, Weighted L^p estimates for powers of selfadjoint operators. Adv. Math. **229**(1), 501–530 (2012)
33. A.-P. Calderón, R. Vaillancourt, On the boundedness of pseudo-differential operators. J. Math. Soc. Jpn. **23**, 374–378 (1971)
34. R.H. Cameron, A family of integrals serving to connect the Wiener and Feynman integrals. J. Math. Phys. **39**, 126–140 (1960)
35. D. Campbell, S. Hencl, F. Konopecký, The weak inverse mapping theorem. Z. Anal. Anwend. **34**(3), 321–342 (2015)
36. E.J. Candès, L. Demanet, The curvelet representation of wave propagators is optimally sparse. Comm. Pure Appl. Math. **58**(11), 1472–1528 (2005)
37. M. Cappiello, J. Toft, Pseudo-differential operators in a Gelfand-Shilov setting. Math. Nachr. **290**(5–6), 738–755 (2017)
38. G. Carron, T. Coulhon, E.-M. Ouhabaz, Gaussian estimates and L^p-boundedness of Riesz means. J. Evol. Equ. **2**(3), 299–317 (2002)
39. P. Cartier, C. DeWitt-Morette, *Functional Integration: Action and Symmetries* (Cambridge University Press, Cambridge, 2006)
40. A. Cauli, F. Nicola, A. Tabacco, Strichartz estimates for the metaplectic representation. Rev. Mat. Iberoamericana **35**(7), 2079–2092 (2019)
41. Y. Choquet-Bruhat, C. DeWitt-Morette, M. Dillard-Bleick, *Analysis, Manifolds and Physics*, 2nd edn. (North-Holland Publishing Co., Amsterdam, 1982)
42. O. Christensen, *An Introduction to Frames and Riesz Bases*, 2nd edn. (Birkhäuser, Basel/Springer, Cham, 2016)
43. L. Cohen, Generalized phase-space distribution functions. J. Math. Phys. **7**, 781–786 (1966)
44. L. Cohen, Time-frequency distributions – a review. Proc. IEEE **77**(7), 941–981 (1989)
45. L. Cohen, *Time-Frequency Analysis* (Prentice Hall PTR, Hoboken, 1995)
46. L. Cohen, *The Weyl operator and its generalization*, vol. 9 (Birkhäuser/Springer Basel AG, Basel, 2013)

47. F. Concetti, J. Toft, Schatten-von Neumann properties for Fourier integral operators with non-smooth symbols. I. Ark. Mat. **47**(2), 295–312 (2009)
48. E. Cordero, F. Nicola, Sharp integral bounds for Wigner distributions. Int. Math. Res. Not. IMRN (6), 1779–1807 (2018)
49. E. Cordero, F. Nicola, Kernel theorems for modulation spaces. J. Fourier Anal. Appl. **25**(1), 131–144 (2019)
50. E. Cordero, L. Rodino, *Time-Frequency Analysis of Operators* (De Gruyter Berlin, Boston, 2020)
51. E. Cordero, S.I. Trapasso, Linear perturbations of the Wigner distribution and the Cohen class. Anal. Appl. (Singap.) **18**(3), 385–422 (2020)
52. E. Cordero, F. Nicola, L. Rodino, Sparsity of Gabor representation of Schrödinger propagators. Appl. Comput. Harmon. Anal. **26**(3), 357–370 (2009)
53. E. Cordero, F. Nicola, L. Rodino, Time-frequency analysis of Fourier integral operators. Commun. Pure Appl. Anal. **9**(1), 1–21 (2010)
54. E. Cordero, K. Gröchenig, F. Nicola, L. Rodino, Wiener algebras of Fourier integral operators. J. Math. Pures Appl. (9) **99**(2), 219–233 (2013)
55. E. Cordero, K. Gröchenig, F. Nicola, L. Rodino, Generalized metaplectic operators and the Schrödinger equation with a potential in the Sjöstrand class. J. Math. Phys. **55** (8), 081506, 17 (2014)
56. E. Cordero, J. Toft, P. Wahlberg, Sharp results for the Weyl product on modulation spaces. J. Funct. Anal. **267**(8), 3016–3057 (2014)
57. E. Cordero, F. Nicola, L. Rodino, Gabor representations of evolution operators. Trans. Amer. Math. Soc. **367**(11), 7639–7663 (2015)
58. E. Cordero, F. Nicola, L. Rodino, Schrödinger equations with rough Hamiltonians. Discrete Contin. Dyn. Syst. **35**(10), 4805–4821 (2015)
59. E. Cordero, F. Nicola, L. Rodino, Wave packet analysis of Schrödinger equations in analytic function spaces. Adv. Math. **278**, 182–209 (2015)
60. E. Cordero, M.A. de Gosson, F. Nicola, On the invertibility of Born-Jordan quantization. J. Math. Pures Appl. (9) **105**(4), 537–557 (2016)
61. E. Cordero, M.A. de Gosson, F. Nicola, Semi-classical time-frequency analysis and applications. Math. Phys. Anal. Geom. **20**(4) (2017). Paper No. 26, 23
62. E. Cordero, M.A. de Gosson, F. Nicola, Time-frequency analysis of Born-Jordan pseudodifferential operators. J. Funct. Anal. **272**(2), 577–598 (2017)
63. E. Cordero, M.A. de Gosson, M. Dörfler, F. Nicola, On the symplectic covariance and interferences of time-frequency distributions. SIAM J. Math. Anal. **50**(2), 2178–2193 (2018)
64. E. Cordero, M.A. de Gosson, F. Nicola, On the reduction of the interferences in the Born-Jordan distribution. Appl. Comput. Harmon. Anal. **44**(2), 230–245 (2018)
65. E. Cordero, L. D'Elia, S.I. Trapasso, Norm estimates for T-pseudodifferential operators in Wiener amalgam and modulation spaces. J. Math. Anal. Appl. **471**(1–2), 541–563 (2019)
66. E. Cordero, F. Nicola, S.I. Trapasso, Almost diagonalization of T-pseudodifferential operators with symbols in Wiener amalgam and modulation spaces. J. Fourier Anal. Appl. **25**(4), 1927–1957 (2019)
67. E. Cordero, F. Nicola, S.I. Trapasso, Dispersion, spreading and sparsity of Gabor wave packets for metaplectic and Schrödinger operators. Appl. Comput. Harmon. Anal. **55**, 405–425 (2021)
68. A. Córdoba, C. Fefferman, Wave packets and Fourier integral operators. Comm. Partial Differ. Equ. **3**(11), 979–1005 (1978)
69. S. Coriasco, M. Ruzhansky, Global L^p continuity of Fourier integral operators. Trans. Amer. Math. Soc. **366**(5), 2575–2596 (2014)
70. P. D'Ancona, F. Nicola, Sharp L^p estimates for Schrödinger groups. Rev. Mat. Iberoamericana **32**(3), 1019–1038 (2016)
71. M.A. de Gosson, *Symplectic Methods in Harmonic Analysis and in Mathematical Physics*, vol. 7 (Birkhäuser/Springer Basel AG, Basel, 2011)
72. M.A. de Gosson, *Born-Jordan Quantization*, vol. 182 (Springer, Cham, 2016)

73. M.A. de Gosson, *The Principles of Newtonian and Quantum Mechanics* (World Scientific Publishing Co. Pte. Ltd., Hackensack, 2017)

74. M.A. de Gosson, Short-time propagators and the Born-Jordan quan- tization rule. Entropy **20**(11) (2018). Paper No. 869, 15

75. P.A.M. Dirac, *The Principles of Quantum Mechanics*, 4th Revised edn. (Oxford University Press, Oxford, 1978)

76. D.L. Donoho, P.B. Stark, Uncertainty principles and signal recovery. SIAM J. Appl. Math. **49**(3), 906–931 (1989)

77. K.-J. Engel, R. Nagel, *A Short Course on Operator Semigroups* (Springer, New York, 2006)

78. M. Esposito, M. Ruzhansky, Pseudo-differential operators with nonlinear quantizing functions. Proc. Roy. Soc. Edinburgh Sect. A **150**(1), 103–130 (2020)

79. L.C. Evans, *Partial Differential Equations*, 2nd edn., vol. 19 (American Mathematical Society, Providence, 2010)

80. A. Fasano, S. Marmi, *Analytical Mechanics* (Oxford University Press, Oxford, 2013)

81. C. Fefferman, E.M. Stein, H^p spaces of several variables. Acta Math. **129**(3–4), 137–193 (1972)

82. H.G. Feichtinger, Banach spaces of distributions of Wiener's type and interpolation, in *Functional Analysis and Approximation (Oberwolfach, 1980)*, vol. 60 (Birkhäuser, Basel, 1981), pp. 153–165

83. H.G. Feichtinger, On a new Segal algebra. Monatsh. Math. **92**(4), 269–289 (1981)

84. H.G. Feichtinger, Banach convolution algebras of Wiener type, in *Functions, Series, Operators, Vol. I, II (Budapest, 1980)*, vol. 35 (North-Holland, Amsterdam, 1983), pp. 509–524

85. H.G. Feichtinger, Generalized amalgams, with applications to Fourier transform. Can. J. Math. **42**(3), 395–409 (1990)

86. H.G. Feichtinger, Modulation spaces on locally compact abelian groups, in *Wavelets and Their Applications*, ed. by S. Thangavelu, M. Krishna, R. Radha (Allied Publishers, New Dehli, 2003), pp. 99–140. Reprint of 1983 technical report, University of Vienna

87. H.G. Feichtinger, Modulation spaces: looking back and ahead. Samp. Theory Signal Image Proces. **5**(2), 109–140 (2006)

88. H.G. Feichtinger, A novel mathematical approach to the theory of translation invariant linear systems, in *Recent Applications of Harmonic Analysis to Function Spaces, Differential Equations, and Data Science* (Birkhäuser, Basel/Springer, Cham, 2017), pp. 483–516

89. H.G. Feichtinger, Classical Fourier analysis via mild distributions. Nonlinear Stud. **26**(4), 783–804 (2019)

90. H.G. Feichtinger, P. Gröbner, Banach spaces of distributions defined by decomposition methods. I. Math. Nachr. **123**, 97–120 (1985)

91. H.G. Feichtinger, K. Gröchenig, Banach spaces related to integrable group representations and their atomic decompositions. I. J. Funct. Anal. **86**(2), 307–340 (1989)

92. H.G. Feichtinger, K. Gröchenig, Banach spaces related to integrable group representations and their atomic decompositions. II. Monatsh. Math. **108**(2–3), 129–148 (1989)

93. H.G. Feichtinger, K. Gröchenig, A unified approach to atomic decompositions via integrable group representations, in *Function Spaces and Applications (Lund, 1986)*, vol. 1302 (Springer, Berlin, 1988), pp. 52–73

94. H.G. Feichtinger, K. Gröchenig, Gabor frames and time-frequency analysis of distributions. J. Funct. Anal. **146**(2), 464–495 (1997)

95. H.G. Feichtinger, M.S. Jakobsen, Distribution theory by Riemann integrals, in *Mathematical Modelling, Optimization, Analytic and Numerical Solutions*, ed. by P. Manchanda, R. Lozi, A. Siddiqi (Springer, Singapore, 2020), pp. 33–76

96. H.G. Feichtinger, T. Strohmer (eds.), *Gabor Analysis and Algorithms: Theory and Applications* (Birkhäuser, Boston, 1998)

97. H.G. Feichtinger, T. Strohmer (eds.), *Advances in Gabor Analysis* (Birkhäuser, Boston, 2002)

98. H.G. Feichtinger, F. Luef, E. Cordero, Banach Gelfand triples for Gabor analysis, in *Pseudo-Differential Operators*, vol. 1949 (Springer, Berlin, 2008), pp. 1–33

99. H.G. Feichtinger, F. Nicola, S.I. Trapasso, On exceptional times for pointwise convergence of integral kernels in Feynman-Trotter path integrals, in *Anomalies in Partial Differential Equations*, vol. 43 (Springer, Cham, 2021), pp. 293–311

100. C. Fernandez, A. Galbis, J. Toft, Convenient descriptions of weight functions in time-frequency analysis (2014). arXiv: 1406.0465

101. D.D.S. Ferreira, W. Staubach, Global and local regularity of Fourier integral operators on weighted and unweighted spaces. Mem. Amer. Math. Soc. **229**(1074), xiv+65 (2014)

102. R.P. Feynman, Space-time approach to non-relativistic quantum mechanics. Rev. Mod. Phys. **20**, 367–387 (1948)

103. R.P. Feynman, Space-time approach to quantum electrodynamics. Phys. Rev. (2) **76**, 769–789 (1949)

104. R.P. Feynman, A.R. Hibbs, *Quantum Mechanics and Path Integrals*, Emended edn. (Dover Publications Inc., Mineola, 2010)

105. P. Flandrin, *Time-Frequency/Time-Scale Analysis*, vol. 10 (Academic Press Inc., San Diego, 1999)

106. G.B. Folland, *Harmonic Analysis in Phase Space*, vol. 122 (Princeton University Press, Princeton, 1989)

107. J.J.F. Fournier, J. Stewart, Amalgams of L^p and l^q. Bull. Amer. Math. Soc. (N.S.) **13**(1), 1–21 (1985)

108. D. Fujiwara, A construction of the fundamental solution for the Schrödinger equation. J. Anal. Math. **35**, 41–96 (1979)

109. D. Fujiwara, Remarks on convergence of the Feynman path integrals. Duke Math. J. **47**(3), 559–600 (1980)

110. D. Fujiwara, Some Feynman path integrals as oscillatory integrals over a Sobolev manifold, in *Functional Analysis and Related Topics, 1991 (Kyoto)*, vol. 1540 (Springer, Berlin, 1993), pp. 39–53

111. D. Fujiwara, *Rigorous Time Slicing Approach to Feynman Path Integrals* (Springer, Tokyo, 2017)

112. D. Fujiwara, N. Kumano-go, Smooth functional derivatives in Feynman path integrals by time slicing approximation. Bull. Sci. Math. **129**(1), 57–79 (2005)

113. D. Gabor, Theory of communication. J. IEE **93**(III), 429–457 (1946)

114. Y.V. Galperin, S. Samarah, Time-frequency analysis on modulation spaces M^{pq} <p q $\le \infty$. Appl. Comput. Harmon. Anal. **16**(1), 1–18 (2004)

115. H. Gask, A proof of Schwartz's kernel theorem. Math. Scand. **8**, 327–332 (1960)

116. R.J. Glauber, Coherent and incoherent states of the radiation field. Phys. Rev. (2) **131**, 2766–2788 (1963)

117. S.S. Goh, T.N.T. Goodman, Estimating maxima of generalized cross ambiguity functions, and uncertainty principles. Appl. Comput. Harmon. Anal. **34**(2), 234–251 (2013)

118. H. Goldstein, *Classical Mechanics*, 2nd edn. (AddisonWesley Publishing Co., Reading, 1980)

119. K. Gröchenig, A pedestrian's approach to pseudodifferential operators, in *Harmonic Analysis and Applications* (Birkhäuser, Boston, 2006), pp. 139–169

120. K. Gröchenig, Composition and spectral invariance of pseudodifferential operators on modulation spaces. J. Anal. Math. **98**, 65–82 (2006)

121. K. Gröchenig, *Foundations of Time-Frequency Analysis* (Birkhäuser Boston Inc., Boston, 2001)

122. K. Gröchenig, Time-frequency analysis of Sjöstrand's class. Rev. Mat. Iberoamericana **22**(2), 703–724 (2006)

123. K. Gröchenig, Weight functions in time-frequency analysis, in *Pseudo-Differential Operators: Partial Differential Equations and Time-Frequency Analysis*, vol. 52 (Amer Mathematical Society, Providence, 2007), pp. 343–366

124. K. Gröchenig, Wiener's lemma: theme and variations. An introduction to spectral invariance and its applications, in *Four Short Courses on Harmonic Analysis* (Birkhäuser, Basel, 2010)

125. K. Gröchenig, Linear independence of time-frequency shifts? Monatsh. Math. **177**(1), 67–77 (2015)
126. K. Gröchenig, Z. Rzeszotnik, Banach algebras of pseudodifferential operators and their almost diagonalization. Ann. Inst. Fourier (Grenoble) **58**(7), 2279–2314 (2008)
127. K. Gröchenig, P. Jaming, E. Malinnikova, Zeros of the Wigner distribution and the short-time Fourier transform. Rev. Mat. Complut. **33**(3), 723–744 (2020)
128. C. Grosche, F. Steiner, *Handbook of Feynman Path Integrals*, vol. 145 (Springer, Berlin, 1998)
129. G. Grubb, *Distributions and Operators*, vol. 252 (Springer, New York, 2009)
130. K. Guo, D. Labate, Sparse shearlet representation of Fourier integral operators. Electron. Res. Announc. Math. Sci. **14**, 7–19 (2007)
131. W. Guo, J. Chen, D. Fan, G. Zhao, Characterizations of some properties on weighted modulation and Wiener amalgam spaces. Michigan Math. J. **68**(3), 451–482 (2019)
132. C. Heil, An introduction to weighted Wiener amalgams, in *Wavelets and Their Applications*, ed. by S. Thangavelu, M. Krishna, R. Radha (Allied Publishers, New Dehli, 2003), pp. 183–216
133. C. Heil, Linear independence of finite Gabor systems, in *Harmonic Analysis and Applications* (Birkhäuser, Boston, 2006), pp. 171–206
134. C. Heil, *A Basis Theory Primer* (Birkhäuser, Basel/Springer, New York, 2011)
135. C. Heil, D. Speegle, The HRT conjecture and the zero divisor conjecture for the Heisenberg group, in *Excursions in Harmonic Analysis*, vol. 3 (Birkhäuser, Basel/Springer, Cham, 2015), pp. 159–176
136. C. Heil, J. Ramanathan, P. Topiwala, Linear independence of time-frequency translates. Proc. Amer. Math. Soc. **124**(9), 2787–2795 (1996)
137. F. Hlawatsch, F. Auger (eds.), *Time-Frequency Analysis: Concepts and Methods* (ISTE, London/John Wiley & Sons, Hoboken, 2008)
138. F. Hlawatsch, G.F. Boudreaux-Bartels, Linear and quadratic time-frequency signal representations. IEEE Signal Process. Mag. **9**(2), 21–67 (1992)
139. F. Holland, Harmonic analysis on amalgams of L^p and f^q. J. Lond. Math. Soc. (2) **10**, 295–305 (1975)
140. A. Holst, J. Toft, P. Wahlberg, Weyl product algebras and modulation spaces. J. Funct. Anal. **251**(2), 463–491 (2007)
141. L. Hörmander, *The Analysis of Linear Partial Differential Operators. III*, vol. 274 (Springer, Berlin, 1985)
142. L. Hörmander, Symplectic classification of quadratic forms, and general Mehler formulas. Math. Z. **219**(3), 413–449 (1995)
143. W. Ichinose, On the formulation of the Feynman path integral through broken line paths. Comm. Math. Phys. **189**(1), 17–33 (1997)
144. W. Ichinose, Convergence of the Feynman path integral in the weighted Sobolev spaces and the representation of correlation functions. J. Math. Soc. Jpn. **55**(4), 957–983 (2003)
145. H. Inci, T. Kappeler, P. Topalov, On the regularity of the composition of diffeomorphisms. Mem. Amer. Math. Soc. **226**(1062), vi+60 (2013)
146. K. Itô, Wiener integral and Feynman integral, in *Proceedings of the Fourth Berkeley Symposium on Mathematical Statistics and Probability*, vol. 2. Contributions to Probability Theory (University of California Press, Berkeley, 1961), pp. 227–238
147. K. Itô, Generalized uniform complex measures in the Hilbertian metric space with their application to the Feynman integral, in *Proceedings of the Fifth Berkeley Symposium on Mathematical Statistics and Probability (Berkeley Calif., 1965/66)*, vol. II. Contributions to Probability Theory Part 1 (Univ California Press, Berkeley, 1967), pp. 145–161
148. M.S. Jakobsen, On a (no longer) new Segal algebra: a review of the Feichtinger algebra. Engl. J. Fourier Anal. Appl. **24**(6), 1579–1660 (2018)
149. A.J.E.M. Janssen, Gabor representation of generalized functions. J. Math. Anal. Appl. **83**(2), 377–394 (1981)

150. A.J.E.M. Janssen, A note on Hudson's theorem about functions with nonnegative Wigner distributions. SIAM J. Math. Anal. **15**(1), 170–176 (1984)
151. A.J.E.M. Janssen, Positivity and spread of bilinear time-frequency distributions, in *The Wigner Distribution* (Elsevier Sci. B. V., Amsterdam, 1997), pp. 1–58
152. A. Jensen, S. Nakamura, Mapping properties of functions of Schrödinger operators between L^p-spaces and Besov spaces, in *Spectral and Scattering Theory and Applications*, vol. 23 (Mathematical Society of Japan, Tokyo, 1994), pp. 187–209
153. A. Jensen, S. Nakamura, L^p-mapping properties of functions of Schrödinger operators and their applications to scattering theory. J. Math. Soc. Jpn. **47**(2), 253–273 (1995)
154. G.W. Johnson, M.L. Lapidus, *The Feynman Integral and Feynman's Operational Calculus* (The Clarendon Press, Oxford University Press, New York, 2000)
155. L. Kapitanski, I. Rodnianski, K. Yajima, On the fundamental solution of a perturbed harmonic oscillator. Topol. Methods Nonlinear Anal. **9**(1), 77–106 (1997)
156. T. Kato, The Cauchy problem for quasi-linear symmetric hyperbolic systems. Arch. Ration. Mech. Anal. **58**(3), 181–205 (1975)
157. J.R. Klauder, *A Modern Approach to Functional Integration* (Birkhäuser, Basel/Springer, New York, 2011)
158. H. Kleinert, *Path Integrals in Quantum Mechanics, Statistics, Polymer Physics, and Financial Markets*, 5th edn. (World Scientific, Singapore, 2009)
159. A.W. Knapp, *Representation Theory of Semisimple Groups* (Princeton University Press, Princeton, 2001)
160. M. Kobayashi, M. Sugimoto, The inclusion relation between Sobolev and modulation spaces. J. Funct. Anal. **260**(11), 3189–3208 (2011)
161. M. Kobayashi, A. Miyachi, N. Tomita, Embedding relations between local Hardy and modulation spaces. Studia Math. **192**(1), 79–96 (2009)
162. H. Koch, D. Tataru, Dispersive estimates for principally normal pseudodifferential operators. Comm. Pure Appl. Math. **58**(2), 217–284 (2005)
163. H. Kumano-go, *Pseudodifferential Operators* (MIT Press, Cambridge, 1981)
164. N. Kumano-go, A construction of the fundamental solution for Schrödinger equations. J. Math. Sci. Univ. Tokyo **2**(2), 441–498 (1995)
165. N. Kumano-go, Feynman path integrals as analysis on path space by time slicing approximation. Bull. Sci. Math. **128**(3), 197–251 (2004)
166. L.D. Landau, E.M. Lifshitz, *Course of Theoretical Physics*, vol. 1, 3rd edn. (Pergamon Press, Oxford, 1976)
167. G. Leoni, *A First Course in Sobolev Spaces*, 2nd edn., vol. 181 (American Mathematical Society, Providence, 2017)
168. Y.I. Lyubarskii, K. Seip, Convergence and summability of Gabor expansions at the Nyquist density. J. Fourier Anal. Appl. **5**(2–3), 127–157 (1999)
169. N. Makri, Feynman path integration in quantum dynamics. Comput. Phys. Comm. **63**(1), 389–414 (1991)
170. N. Makri, W.H. Miller, Correct short time propagator for Feynman path integration by power series expansion in $1't$. Chem. Phys. Lett. **151**, 1–8 (1988)
171. N. Makri, W.H. Miller, Exponential power series expansion for the quantum time evolution operator. J. Chem. Phys. **90**(2), 904–911 (1988)
172. J. Marzuola, J. Metcalfe, D. Tataru, Wave packet parametrices for evolutions governed by PDO's with rough symbols. Proc. Amer. Math. Soc. **136**(2), 597–604 (2008)
173. V.P. Maslov, *Théorie des perturbations et méthodes asymptotiques*, French translation from Russian edn. (Dunod, Paris, 1970)
174. S. Mazzucchi, *Mathematical Feynman Path Integrals and Their Applications* (World Scientific Publishing Co. Pte. Ltd., Hackensack, 2009)
175. W. Mecklenbräuker, F. Hlawatsch (eds.), *The Wigner Distribution* (Elsevier Science B.V., Amsterdam, 1997)
176. R.B. Melrose, *Geometric Scattering Theory* (Cambridge University Press, Cambridge, 1995)

177. A. Miyachi, On some Fourier multipliers for $H^p(\mathbf{R}^n)$. J. Fac. Sci. Univ Tokyo Sect. IA Math. **27**(1), 157–179 (1980)

178. E. Nelson, Feynman integrals and the Schrödinger equation. J. Math. Phys. **5**, 332–343 (1964)

179. F. Nicola, Convergence in L^p for Feynman path integrals. Adv. Math. **294**, 384–409 (2016)

180. F. Nicola, On the time slicing approximation of Feynman path integrals for non-smooth potentials. J. Anal. Math. **137**(2), 529–558 (2019)

181. F. Nicola, L. Rodino, *Global Pseudo-Differential Calculus on Euclidean Spaces*, vol. 4 (Birkhäuser Verlag, Basel, 2010)

182. F. Nicola, S.I. Trapasso, Approximation of Feynman path integrals with non-smooth potentials. J. Math. Phys. **60**(10), 102103, 13 (2019)

183. F. Nicola, S.I. Trapasso, On the pointwise convergence of the integral kernels in the Feynman-Trotter formula. Comm. Math. Phys. **376**(3), 2277–2299 (2020)

184. L. Nirenberg, On elliptic partial differential equations. Ann. Scuola Norm. Sup. Pisa Cl. Sci. (3) **13**, 115–162 (1959)

185. K.A. Okoudjou, Extension and restriction principles for the HRT conjecture. J. Fourier Anal. Appl. **25**(4), 1874–1901 (2019)

186. J. Peetre, *New Thoughts on Besov Spaces*. Mathematics Department (Duke University, Durham, 1976)

187. A.M. Perelomov, Remark on the completeness of the coherent state system. Teor. Mat. Fiz. **6**(2), 213–224 (1971)

188. A.M. Perelomov, *Generalized Coherent States and Their Applications* (Springer, Berlin, 1986)

189. S. Pilipović, N. Teofanov, Pseudodifferential operators on ultra-modulation spaces. J. Funct. Anal. **208**(1), 194–228 (2004)

190. M. Reed, B. Simon, *Methods of Modern Mathematical Physics*. 1. Functional analysis (Academic Press, New York, 1972)

191. M. Reed, B. Simon, *Methods of Modern Mathematical Physics*. 1. Fourier analysis, self-adjointness (Academic Press [Harcourt Brace Jovanovich, Publishers], New York, 1975)

192. M. Reich, W. Sickel, Multiplication and composition in weighted modulation spaces, in *Mathematical Analysis, Probability and Applications – Plenary Lectures*, vol. 177 (Springer, Cham, 2016), pp. 103–149

193. J. Robbin, D. Salamon, Feynman path integrals on phase space and the metaplectic representation. Math. Z. **221**(2), 307–335 (1996)

194. R. Rochberg, K. Tachizawa, Pseudodifferential operators, Gabor frames, and local trigonometric bases, in *Gabor Analysis and Algorithms* (Birkhäuser, Boston, 1998), pp. 171–192

195. W. Rudin, *Functional Analysis*, 2nd edn. (McGraw-Hill Inc., New York, 1991)

196. M. Ruzhansky, *Regularity Theory of Fourier Integral Operators with Complex Phases and Singularities of Affine Fibrations*, vol. 131 (Stichting Mathematisch Centrum, Centrum voor Wiskunde en Informatica, Amsterdam, 2001)

197. M. Ruzhansky, M. Sugimoto, Global L^2-boundedness theorems for a class of Fourier integral operators. Comm. Partial Differ. Equ. **31**(4–6), 547–569 (2006)

198. M. Ruzhansky, M. Sugimoto, B. Wang, Modulation spaces and nonlinear evolution equations, in *Evolution Equations of Hyperbolic and Schrödinger Type*, vol. 301 (Birkhäuser, Basel/Springer, Basel, 2012), pp. 267–283

199. T. Sauer, Remarks on the origin of path integration: Einstein and Feynman, in *Path Integrals* (World Sci. Publ., Hackensack, 2008), pp. 3–13

200. E. Schrödinger, Der stetige Übergang von der Mikro- zur Makromechanik, in *Collected Papers on Wave Mechanics*. Translated from the second German edition by J.F. Shearer, W.M. Deans, Including ıt Four lectures on wave mechanics (Chelsea Publishing Co., New York, 1982), p. xiii+207

201. L.S. Schulman, *Techniques and Applications of Path Integration* (John Wiley & Sons, Inc., New York, 1981)

202. L. Schwartz, *Théorie des Distributions* (Hermann, Paris, 1966)

203. A. Seeger, C.D. Sogge, E.M. Stein, Regularity properties of Fourier integral operators Ann. Math. (2) **134**(2), 231–251 (1991)
204. K. Seip, Density theorems for sampling and interpolation in the Bargmann-Fock space. I. J. Reine Angew Math. **429**, 91–106 (1992)
205. K. Seip, R. Wallstén, Density theorems for sampling and interpolation in the Bargmann-Fock space. II. J. Reine Angew. Math. **429**, 107–113 (1992)
206. A. Serafini, *Quantum Continuous Variables* (CRC Press, Boca Raton, 2017)
207. M.A. Shubin, *Pseudodifferential Operators and Spectral Theory* (Springer, Berlin, 1987)
208. B. Simon, *Functional Integration and Quantum Physics*, 2nd edn. (AMS Chelsea Publishing, Providence, 2005)
209. J. Sjöstrand, An algebra of pseudodifferential operators. Math. Res. Lett. **1**(2), 185–192 (1994)
210. H.F. Smith, D. Tataru, Sharp local well-posedness results for the nonlinear wave equation. Ann. Math. (2) **162**(1), 291–366 (2005)
211. E.M. Stein, *Singular Integrals and Differentiability Properties of Functions* (Princeton University Press, Princeton, 1970)
212. E.M. Stein, *Harmonic Analysis: Real-Variable Methods, Orthogonality and Oscillatory Integrals*, vol. 43 (Princeton University Press, Princeton, 1993)
213. T. Tao, *Nonlinear Dispersive Equations: Local and Global Analysis* (Amer Mathematical Society, Providence, 2006)
214. D. Tataru, Phase space transforms and microlocal analysis, in *Phase Space Analysis of Partial Differential Equations*, vol. II (Scuola Normale Superiore, Pisa, 2004), pp. 505–524
215. M.E. Taylor, *Partial Differential Equations II. Qualitative Studies of Linear Equations*, 2nd edn., vol. 116 (Springer, New York, 2011)
216. N. Teofanov, Ultradistributions and time-frequency analysis, in *Pseudo-Differential Operators and Related Topics*, vol. 164 (Birkhäuser, Basel, 2006), pp. 173–192
217. J. Toft, Continuity properties for modulation spaces, with applications to pseudo-differential calculus. I J. Funct. Anal. **207**(2), 399–429 (2004)
218. J. Toft, Continuity properties for modulation spaces, with applications to pseudo-differential calculus. II. Ann. Glob. Anal. Geom. **26**(1), 73–106 (2004)
219. J. Toft, Matrix parameterized pseudo-differential calculi on modulation spaces, in *Generalized Functions and Fourier Analysis*, vol. 260 (Birkhäuser, Basel/Springer, Cham, 2017), pp. 215–235
220. J. Toft, F. Concetti, G. Garello, Schatten-von Neumann properties for Fourier integral operators with non-smooth symbols II. Osaka J. Math. **47**(3), 739–786 (2010)
221. J. Toft, K. Johansson, S. Pilipović, N. Teofanov, Sharp convolution and multiplication estimates in weighted spaces. Anal. Appl. (Singap.) **13**(5), 457–480 (2015)
222. S.I. Trapasso, Time-frequency analysis of the Dirac equation. J. Differ. Equ. **269**(3), 2477–2502 (2020)
223. F. Treves, *Topological Vector Spaces, Distributions and Kernels* (Academic Press, New York, 1967)
224. H. Triebel, *Interpolation Theory Function Spaces, Differential Operators*, vol. 18 (North-Holland Publishing Co., Amsterdam, 1978)
225. H. Triebel, *Theory of Function Spaces. II*, vol. 84 (Birkhäuser, Basel, 1992)
226. T. Tsuchida, Remarks on Fujiwara's stationary phase method on a space of large dimension with a phase function involving electromagnetic fields. Nagoya Math. J. **136**, 157–189 (1994)
227. V. Turunen, Born-Jordan time-frequency analysis, in *Harmonic Analysis and Nonlinear Partial Differential Equations* (Research Institute for Mathematical Sciences (RIMS), Kyoto, 2016), pp. 107–186
228. J. Ville, Théorie et applications de la notion de signal analytique. Câbles et Transm. **2A**, 61–74 (1948)
229. J. von Neumann, *Mathematical Foundations of Quantum Mechanics* (Princeton University Press, Princeton, 2018)

230. P. Wahlberg, Vectorvalued modulation spaces and localization operators with operatorvalued symbols. Integr. Equ. Oper. Theory **59**(1), 99–128 (2007)

231. B. Wang, L. Zhao, B. Guo, Isometric decomposition operators, function spaces E^{ll} and applications to nonlinear evolution equations. J. Funct. Anal. **233**(1), 1–39 (2006)

232. B. Wang, Z. Huo, C. Hao, Z. Guo, *Har Monic Analysis Method for Nonlinear Evolution Equations. I* (World Scientific Publishing Co. Pte. Ltd., Hackensack, 2011)

233. A. Weinstein, A symbol class for some Schrödinger equations on $\mathbf{R}^n$. Amer. J. Math. **107**(1), 1–21 (1985)

234. H. Weyl, *The Theory of Groups and Quantum Mechanics* (Dover Publications Inc., New York, 2014)

235. N. Wiener, On the representation of functions by trigonometrical integrals. Math. Z. **24**(1), 575–616 (1926)

236. N. Wiener, Tauberian theorems. Ann. Math. (2) **33**(1), 1–100 (1932)

237. N. Wiener, *The Fourier Integral and Certain of Its Applications* (Cambridge University Press, Cambridge, 1988)

238. E. Wigner, On the quantum correction for thermodynamic equilibrium. Phys. Rev. **40**(5), 749–759 (1932)

239. T.H. Wolff, *Lectures on Harmonic Analysis*, vol. 29 (American Mathematical Society, Providence, 2003)

240. M.W. Wong, *Weyl Transforms* (Springer, New York, 1998)

241. K. Yajima, Schrödinger evolution equations with magnetic fields. J. Anal. Math. **56**, 29–76 (1991)

242. M. Zworski, *Semiclassical Analysis*, vol. 138 (American Mathematical Society, Providence, 2012)

Index

© The Author(s), under exclusive license to Springer Nature Switzerland AG 2022
F. Nicola, S. I. Trapasso, *Wave Packet Analysis of Feynman Path Integrals*,
Lecture Notes in Mathematics 2305, https://doi.org/10.1007/978-3-031-06186-8

LECTURE NOTES IN MATHEMATICS

Editors in Chief: J.-M. Morel, B. Teissier;

Editorial Policy

1. Lecture Notes aim to report new developments in all areas of mathematics and their applications – quickly, informally and at a high level. Mathematical texts analysing new developments in modelling and numerical simulation are welcome.

   Manuscripts should be reasonably self-contained and rounded off. Thus they may, and often will, present not only results of the author but also related work by other people. They may be based on specialised lecture courses. Furthermore, the manuscripts should provide sufficient motivation, examples and applications. This clearly distinguishes Lecture Notes from journal articles or technical reports which normally are very concise. Articles intended for a journal but too long to be accepted by most journals, usually do not have this "lecture notes" character. For similar reasons it is unusual for doctoral theses to be accepted for the Lecture Notes series, though habilitation theses may be appropriate.

2. Besides monographs, multi-author manuscripts resulting from SUMMER SCHOOLS or similar INTENSIVE COURSES are welcome, provided their objective was held to present an active mathematical topic to an audience at the beginning or intermediate graduate level (a list of participants should be provided).

   The resulting manuscript should not be just a collection of course notes, but should require advance planning and coordination among the main lecturers. The subject matter should dictate the structure of the book. This structure should be motivated and explained in a scientific introduction, and the notation, references, index and formulation of results should be, if possible, unified by the editors. Each contribution should have an abstract and an introduction referring to the other contributions. In other words, more preparatory work must go into a multi-authored volume than simply assembling a disparate collection of papers, communicated at the event.

3. Manuscripts should be submitted either online at www.editorialmanager.com/lnm to Springer's mathematics editorial in Heidelberg, or electronically to one of the series editors. Authors should be aware that incomplete or insufficiently close-to-final manuscripts almost always result in longer refereeing times and nevertheless unclear referees' recommendations, making further refereeing of a final draft necessary. The strict minimum amount of material that will be considered should include a detailed outline describing the planned contents of each chapter, a bibliography and several sample chapters. Parallel submission of a manuscript to another publisher while under consideration for LNM is not acceptable and can lead to rejection.

4. In general, **monographs** will be sent out to at least 2 external referees for evaluation.

   A final decision to publish can be made only on the basis of the complete manuscript, however a refereeing process leading to a preliminary decision can be based on a pre-final or incomplete manuscript.

   Volume Editors of **multi-author works** are expected to arrange for the refereeing, to the usual scientific standards, of the individual contributions. If the resulting reports can be forwarded to the LNM Editorial Board, this is very helpful. If no reports are forwarded or if other questions remain unclear in respect of homogeneity etc, the series editors may wish to consult external referees for an overall evaluation of the volume.

5. Manuscripts should in general be submitted in English. Final manuscripts should contain at least 100 pages of mathematical text and should always include

   - a table of contents;
   - an informative introduction, with adequate motivation and perhaps some historical remarks: it should be accessible to a reader not intimately familiar with the topic treated;
   - a subject index: as a rule this is genuinely helpful for the reader.
   - For evaluation purposes, manuscripts should be submitted as pdf files.

6. Careful preparation of the manuscripts will help keep production time short besides ensuring satisfactory appearance of the finished book in print and online. After acceptance of the manuscript authors will be asked to prepare the final LaTeX source files (see LaTeX templates online: https://www.springer.com/gb/authors-editors/book-authors-editors/manuscriptpreparation/5636) plus the corresponding pdf- or zipped ps-file. The LaTeX source files are essential for producing the full-text online version of the book, see http://link.springer.com/bookseries/304 for the existing online volumes of LNM). The technical production of a Lecture Notes volume takes approximately 12 weeks. Additional instructions, if necessary, are available on request from lnm@springer.com.

7. Authors receive a total of 30 free copies of their volume and free access to their book on SpringerLink, but no royalties. They are entitled to a discount of 33.3 % on the price of Springer books purchased for their personal use, if ordering directly from Springer.

8. Commitment to publish is made by a *Publishing Agreement*; contributing authors of multiauthor books are requested to sign a *Consent to Publish form*. Springer-Verlag registers the copyright for each volume. Authors are free to reuse material contained in their LNM volumes in later publications: a brief written (or e-mail) request for formal permission is sufficient.

Addresses:
Professor Jean-Michel Morel, CMLA, École Normale Supérieure de Cachan, France
E-mail: moreljeanmichel@gmail.com

Professor Bernard Teissier, Equipe Géométrie et Dynamique,
Institut de Mathématiques de Jussieu – Paris Rive Gauche, Paris, France
E-mail: bernard.teissier@imj-prg.fr

Springer: Ute McCrory, Mathematics, Heidelberg, Germany,
E-mail: lnm@springer.com

Printed in the United States
by Baker & Taylor Publisher Services